Backyard Bigfoot

BACKYARD BIGFOOT

The True Story of Stick Signs, UFOs, & the Sasquatch

LISA A. SHIEL

with a foreword by Nick Redfern

Slipdown Mountain Publications LLC
Lake Linden, MI
Toll-Free: 1-877-388-5227

BACKYARD BIGFOOT
BY LISA A. SHIEL

First Edition

Manufactured in the United States

ISBN 13: 978-0-9746553-6-9
ISBN 10: 0-9746553-6-8

Library of Congress Control Number: 2005908829

Information: www.SlipdownMountain.com
1-877-388-5227

SLIPDOWN MOUNTAIN PUBLICATIONS

Publisher's Cataloging-in-Publication Data

Shiel, Lisa A.

Backyard Bigfoot : the true story of stick signs, UFOs, & the Sasquatch / Lisa A. Shiel ; with a foreword by Nick Redfern.

p. cm.

1. Sasquatch. 2. Hominids—Evolution. 3. Unidentified flying objects.

QL89.2.H54 B33 2006
001.9'44—dc22 2005908829

To "the foot"

Contents

Illustrations

Chapter 1
Ancient Evidence

Chapter 2
Stick Signs

Chapter 3
Dubious Origins

Chapter 4
Mistaken Identity

Chapter 5
Orbs in the Night

Chapter 6
Tying the Knots

Foreword

by Nick Redfern

Countless words have been written on that most mysterious of all man-beasts, Bigfoot, and doubtless countless more will continue to be written until the ever-vexing question of what the beast is (or indeed is not!) is finally resolved to the satisfaction of the collective cryptozoological and scientific research communities.

Until that day arrives, we are largely reliant upon the work of investigators that dedicate their lives to studying and chronicling the facts surrounding all things large, lumbering and hairy. In addition, securing the testimony of firsthand witnesses to Bigfoot and its handiwork is overwhelmingly crucial when it comes to keeping in touch with the latest developments, sightings, encounters, and activities of the monsters that lurk deep within the shadowy woods and the darkest corners of North America.

Seldom, however, is the researcher, the writer and the witness the same person. And seldom is that same person able to put their thoughts, experiences, discoveries and conclusions into book format in an insightful, thought-provoking and always-entertaining manner. But such is the nature of Lisa Shiel and her new book, *Backyard Bigfoot: The True Story Of Stick Signs, UFOs, And The Sasquatch.*

Perhaps best known up until now for her entertaining novels *The Hunt for Bigfoot* and *Lord of the Dead,* Lisa now turns her hand to nonfiction—and not before time: anyone wanting a unique, and intensely personal perspective on the Bigfoot controversy will not be disappointed by *Backyard*

Bigfoot. And Lisa is to be applauded for addressing highly controversial issues that many within cryptozoology are steadfastly determined to shy away from when it comes to Bigfoot—particularly the controversies surrounding UFOs and Bigfoot, and the personal interaction between their species and ours.

Lisa's book begins with a superb roundup of reports of hairy man-beasts dating back thousands of years and across numerous continents, a plethora of cultures, and amid a range of ancient legends, tales, and assorted folklore. And after reading this opening salvo of data, it becomes abundantly clear that Bigfoot-style entities are not just with us today, but have been for perhaps as long as we ourselves have existed.

But it is in Chapter Two that the adventure and the intrigue really begins, when Lisa introduces us to her personal research into a seldom-discussed aspect of the Bigfoot puzzle: stick-signs. Many skeptics—and indeed researchers—bemoan the lack of hard evidence indicating a Bigfoot presence among us. But if you know where to look, you can find it. For example, in mid-2005, myself and well-known Texas monster hunter Ken Gerhard and his wife, Lori, uncovered a classic teepee-style creation while roaming the woods of the Lone Star State's Lake Worth, in search of the infamous Goat Man. Teepees constructed out of branches, as many students of Bigfoot will be aware, are found with regular frequency in areas of man-beast activity. But they are not the only things found, as Lisa reveals.

At a personal level on this aspect of the Bigfoot mystery, things really kicked off for Lisa in November 2003 while she was living in the woodlands of North Texas. She was already aware of what appeared to be unusual presences in the vicinity of her home during the dead of night—presences that would spook both her pet animals and Lisa, too. But it was when *something* began to leave carefully placed sticks on the ground of her remote property in clearly delineated patterns that things moved up a gear. I will not relate the full story of what happened, as Lisa expertly does so herself. But I will say this: if you want to learn more about the way in which this mysterious creature (whatever its point of origin) interacts with the human species, then Lisa's revelations and intriguing experiences are essential reading.

Indeed, as Lisa shows, what we appear to have among us is not some form of unintelligent ape; but rather a creature or entity that we have vastly underestimated and that possesses skills and powers far beyond our own and that displays curiosity, playfulness, and, perhaps, even humor.

Controversies such as the conventional picture of Darwin's Theory of Evolution, the rise of ancient civilizations, and Earth's recent and ancient history are discussed, dissected and shown to be highly problematic—as is the theory subscribed to by a number of researchers that Bigfoot is nothing more than the last remnants of that most massive of all apes—*Gigantopithecus*. Some Bigfoot scholars will not agree with Lisa's conclusions on this latter issue. So be it. But no one can say that Lisa does not firmly stand behind her assertions—and for that alone she deserves our admiration and support.

Chapter Five, Orbs in the Night, will doubtless stir up those that adhere to the idea that Bigfoot is purely a zoological problem, primarily because it addresses the various reports and data that suggests some form of link between Bigfoot and the equally controversial UFO puzzle. But anyone and everyone that claims to be open-minded on the Bigfoot topic should read this data very carefully. Something very strange is indeed afoot.

Synchronicities and weird coincidences also play a role in this story when the critters that Lisa is seeking out in Texas, resurface when she moves to Michigan to live. Again, I will let you, the reader, savor the delights of this aspect of the story, but, once more, it is a fascinating addition to the world of Bigfoot.

Backyard Bigfoot is manifestly different to so many other titles on the bookshelves, and I am mightily relieved and pleased that it is. Reference-style books are fine; but if you are determined to go to the next level of your understanding and appreciation of the creature that is Bigfoot, listen to the witnesses. But when the author and the witness are one and the same, you are in for a real treat.

Nick Redfern is the author of *Three Men Seeking Monsters: Six Weeks In Pursuit Of Werewolves, Lake Monsters, Giant Cats, Ghostly Devil Dogs, And Ape-Men* **(Simon & Schuster, 2004); and** *On The Trail Of The Saucer Spies* **(Anomalist Books, 2006).**

Introduction

As the first humans trundled out of Africa 200,000 years ago, a giant ape known as *Gigantopithecus* roamed southeast Asia. Today, the descendants of both species endure—us and Bigfoot. This is the story told by most cryptozoologists. But is the story accurate? Or have both archaeologists and cryptozoologists ignored vital evidence in their quests to prove their theories?

This book presents evidence that suggests they have, and that our view of human and Bigfoot origins has been skewed to support the unproven theory of evolution. Only by reexamining the evidence for human evolution, in the light cast by forgotten evidence, can we understand the true nature of Bigfoot.

These hairy, bipedal creatures can come in every shape and size imaginable, from giants to pygmies, and apelike to humanoid. The name "Bigfoot" applies to one variant of the species, whose large feet and giant stature inspired the nickname. Conversely, the umbrella term "hairy hominid," a label saddled to neither one physical attribute nor B-movie overtones, describes the species better while encompassing the full range of variations among the creatures. Hominid refers to any species of humanlike, bipedal (walking on two feet) creature. The term has come to stand for any biped presumed to be an ancestor of humans, though I will use the word in its more general sense. The hominid genre includes *Australopithecus, Homo erectus,* and the Neanderthals. Hairy hominid, then, denotes a hairy, humanlike biped.

UFOs, hairy dwarves, glowing-eyed monsters, mysterious footprints—these phenomena and others play a role in the hairy hominid mystery. Cryptozoologists prefer to ignore the bizarre sightings, sweeping them into a dark corner where the so-called "lunatic fringe" hides. On most email discussion groups devoted to hairy hominids, any mention of UFOs or other strangeness triggers a furor of criticism that silences the issue. "Bigfoot is a giant ape, nothing more," the critics charge. Yet the evidence—ancient and modern, archaeological and folkloric—belies their assertions. The irony is that proponents of the "big, dumb ape" theory employ the same tactics as the debunkers who refute the existence of hairy hominids: if evidence contradicts your theory, ridicule it until it goes away.

I have chosen to write a book about hairy hominids for a number of reasons, two of which stand out above the rest. Firstly, while many researchers have examined the aboriginal legends of North America and Australia, few have delved deeply enough into the pool of ancient evidence to see what lies beneath the surface. In researching my novel *The Hunt for Bigfoot*, I scoured the archives of ancient knowledge to uncover little-known facts about human history and our relationship to hairy hominids, a kinship that goes back at least as far as the last ice age. Secondly, only a handful of American researchers have investigated the connection between hairy hominids and UFOs. As I will demonstrate later, that connection dates back thousands of years and has much to tell us about our own origins.

As I wrote this book, a new sighting occurred north of Winnipeg in the Canadian province of Manitoba. On April 16, 2005, a ferry-boat operator living on the Norway House Cree Nation reservation caught on video a dark-colored creature that walked upright across the far shore of Lake Winnipeg. The sighting ignited a bonfire of attention around the witness, Bobby Clarke, who became a bit of a recluse due to the overwhelming attentions of the media.

On April 20, four days after the sighting, talk-radio host Glenn Pelletier of Toronto's 570 News asked me to appear on his show to discuss the Bigfoot phenomenon in light of the new sighting. At that point, no one had seen the video, save for a smattering of people who had paraded through the Clarke household to view the footage. Clarke's brother also appeared on Pelletier's show to explain what his brother saw and how it affected him. According to the brother, Clarke was so shaken by the incident that immediately afterward he took his ferry boat out into the middle of the lake, where he stayed for the remainder of the day.

A bidding war ensued for the rights to the footage. Two weeks later, Clarke licensed (as opposed to selling outright) the video to an American tabloid show called *A Current Affair*, for an undisclosed sum. Speculation ran rampant that Clarke had netted six figures for the rights to his video, but no one knows for sure how much the show paid. In the meantime, the Canadian media scrambled to keep the story afloat. They even went so far as to find a zoologist who would debunk the video, sight unseen. One article quoted a Canadian Bigfoot researcher as saying the video was a hoax, though he made no mention of actually having seen the footage.

On May 4, amid much hype, *A Current Affair* debuted the footage. Although Clarke had filmed the creature for nearly three minutes, the show chose to air only 20-30 seconds of the footage. The image was extremely blurry, with the whatever-it-was tromping around about 750 feet from the camera. All the video demonstrated for certain was that Clarke had seen a dark-colored biped—clearly not a bear, as arms and legs are intermittently visible. A reporter from *The Globe and Mail*, Toronto's premier newspaper, asked for my thoughts on the video that night. As I told him, the video is intriguing but inconclusive. Perhaps further analysis can tease more details from the footage.

After the footage aired, *A Current Affair* sent a camera crew to Manitoba to "investigate." Soon the show held a press conference in which the expedition members claimed to have discovered a "Bigfoot feeding ground"—whatever that means! No actual evidence resulted from the expedition.

Just a few months later, in July, nine residents of Teslin in Canada's Yukon Territory reported seeing a similar creature and finding a set of huge footprints.[3] This time, however, some of the witnesses claimed to have gathered hair samples left behind by the creature. Hype and overexcitement followed as the media and Bigfoot enthusiasts awaited results of scientific testing on the hairs. When the University of Alberta announced the results, you could virtually hear the bubble bursting. The hairs belonged to a bison—moreover, the scientist who performed the tests said the hairs were either old or came from tanned hide.[4] The test results washed away a more interesting fact: residents of Teslin had seen something, and found footprints belonging to a large creature. As usual, media attention served to obscure the real encounters rather than focusing public attention on them.

The same scenario plays out after every publicized sighting. The media besieges the witness, in person or by phone, as do a handful of the more aggressive researchers. The hairy hominid who started the uproar has long since departed, or gone into hiding, while the beleaguered witness tires of the spotlight and the ridicule from mainstream scientists. Rarely do either the researchers or the media find more evidence from the same site.

Mere months before Clarke's sighting snagged publicity, an equally compelling story failed to generate interest, probably because the tidbit lay buried in an article about historical documents from the English Civil War. In a December 2004 article, the newspaper The Scotsman reported on the discovery of a trove of forgotten documents from the 17th Century found inside government offices in the UK. The story was titled "English Civil War Documents Unearthed." You'd hardly expect to find hairy hominids in such a report.

But you will. The story describes the 900 pamphlets, dating from the mid-17th Century, which mostly relate to King Charles I. About halfway through the article, however, a set of pamphlets with the title "Wonders at Home and Abroad" is mentioned. According to the article, the pamphlets describe expeditions sent out around the world to search for unusual creatures. On the South Pacific islands of Benganga and Coma, explorers encountered giant, hairy natives who traveled in canoes. The shortest native, the pamphlet says, stood over 11 feet tall.[47] I reported the story in my defunct e-newsletter, Origins E-News. As far as I could determine, no other Bigfoot researchers noticed the article.

By focusing on my own backyard, both literally and figuratively, I gain a long-term perspective on the hairy hominid phenomenon. I don't claim to have all the answers, or even definitive answers. I can only present the fruits of my endeavors, along with my thoughts on them.

In Chapter Two, I discuss the most unusual part of my research—my first-hand experiences with stick signs. Other researchers have used the term "stick signs" to refer to small limbs broken or twisted off trees. When I talk about sticks signs, however, I mean signs constructed out of sticks. They lie on the ground, are easy to overlook, and provide an intriguing glimpse into the psyche of hairy hominids. I'll thrash out the arguments cynics have made in debunking stick signs, as well as explaining what I've learned from my experiences with this unique form of communication.

In my quest to unmask the true identity of hairy hominids, I spent over four years researching human origins and evolution, two subjects intertwined with both each other and the origins of hairy hominids. In this book, I scrutinize the genetic and fossil evidence that led me to draft my own theories about the identity and nature of hairy hominids—as well as humans.

Debunkers argue that sightings prove nothing; eyewitness testimony is unreliable. Perhaps the skeptics should talk to the judges and juries who have convicted defendants based on eyewitness testimony. Yes, eyewitnesses can get confused or misidentify ordinary objects. But the same cynics who decry the validity of eyewitness testimony believe, without reservation or inquiry, the claims of any alleged hoaxers. The annals of both UFO and hairy hominid sightings include thousands of rational, intelligent people who sought ordinary explanations for the extraordinary phenomena they witnessed, but failed to find one. I count myself among those people.

I've based this book on my own extensive research, first-hand encounters, and avid interest in all aspects of the phenomenon. Until a body is found, however—and perhaps even then—the subject of hairy hominids won't make it into a science textbook. Instead of focusing on "proving" they exist, we should concentrate on figuring out what they are, why they're here, and what those answers tell us about ourselves.

The hunt for truth never ends.

1
Ancient Evidence

We humans have long struggled to understand our world. From natural processes like earthquakes to the staggering array of living things that surrounds us, we seek to document all that we experience. The earliest humans, who lacked the technology to sequence DNA or examine tissue samples under a microscope, sought to document and understand their world through the arts. Cave paintings, rock art, sculpture, myths—these served as the tools for ancient archaeologists.

No modern human can claim to have deciphered all the riddles inherent in ancient artwork and myths. So archaeologists have devised an unwritten rule concerning ancient cultures: if you can't explain the symbolism, call it a religious icon. If we remove the lens through which archaeologists and anthropologists have forced us to view the ancient world, we can study the evidence without distortion and gain a sharper understanding of what our ancestors saw. If we treat the ancients as intelligent people rather than superstitious dolts, if we take what they depicted as literally as possible, we may see things we're not supposed to see.

We might see hairy hominids.

Into the Ice Age

The earliest artwork dates to a period called the Paleolithic—literally, "old stone"—which we refer to as the Stone Age. Historians divide the

Paleolithic into three segments labeled Lower, Middle, and Upper; the Lower Paleolithic is the oldest portion. The Paleolithic began about 4.0 million years ago and ended at different times in each region of the world. In fact, some cultures technically live in the Paleolithic today, as they continue to use stone tools similar to those in use thousands of years ago. In most areas, such as Europe and the Mediterranean rim, the stone age concluded around 10,000 years ago.

Science tells us the only "humans" alive during the Lower Paleolithic, 4.0 million years ago, were apelike hominids such as *Australopithecus, Homo erectus,* or the Neanderthals. Science also tells us modern humans emerged in Africa around 200,000 years ago, at the start of the Middle Paleolithic. By the time the ice sheets had begun to retreat, modern humans had found a toehold in Europe and Asia and the earliest Australians had braved the ocean to reach their new homeland. In the Upper Paleolithic, we discover the first traces of culture in the form of rock art, engraved bones, and figurines.

In the Upper Paleolithic, which began 30,000-40,000 years ago, ice sheets swathed much of the planet. Parts of the world that have since turned to desert, such as Egypt, existed as savannahs and lush grasslands. The climate was changing, however, and by the end of the ice age—as the glaciers melted at breakneck speed—the sea level would rise 400 feet.[34] Some researchers believe the rise happened suddenly in a global deluge that inspired the flood myths abundant in cultures around the world. Severe, inexplicable swings in climate triggered animal extinctions during the same period.[34] Humans alive at the time struggled to keep up with the shifting landscape. Species once hunted for food, like the mammoth, died out. Humans had to adapt. Is it coincidence the Paleolithic terminated at the peak of the glacial retreat?

Yet, amid the chaos, humans found time to express themselves in art. From cave paintings to petroglyphs to portable objects, artwork from the Upper Paleolithic astounds us with its sophistication. The artists depicted animals familiar to them, including bison and horses, as well as bizarre beings and abstract symbols whose meaning remains unclear. In these earliest artworks, we can also find images of hairy, bipedal creatures.

The Dolní Head

A site called Dolní Vestonice, in the Czech Republic, stands as one of the most important sites for Paleolithic archaeology. Huts and burials

The ivory head found in the 19th Century (Dolní Vestonice, Czech Republic)—24,000 BC.

uncovered at the site date back to about 28,000 years ago. One hut even contained an oven which the inhabitants had used to fire clay figurines. Archaeologists have found numerous works of art at the site. But one sculpture, offered to the British Museum in the 1940s, affords the most startling glimpse into ice–age life.

The mammoth–ivory sculpture, a head broken off a larger and now lost figure, measures three inches tall. The head depicts a male with a heavy brow ridge, deep–set eyes, wide mouth and nose, stocky neck, and long hair. The hair, which seems to drape around his neck, appears plastered to his skin.

Did the Paleolithic artist attempt to mimic the appearance of a hairy hominid? The creatures are often described as having longer hair on their heads and shoulders. The protruding brow hardly resembles Paleolithic people, for the humans of the ice age looked just like modern people.

One sect of archaeologists dismisses the sculpture as a forgery, for three main reasons:

1. No archaeologist took part in its discovery. A Czech family gave the sculpture to archaeologist Alexander Marshak, who wrote an article about the piece for *National Geographic* in 1988. The family said the piece had been discovered in the 1800s. According to the archaeological elite, the object therefore has no "provenance"—no proven origin—and no context from which to derive a date.
2. The style is too sophisticated and realistic, making the piece unique—a dirty word in archaeology. Paleolithic people either lacked the expertise to carve realistic figures or they preferred stylized forms.
3. The head doesn't look like a modern human. Instead, the male face resembles a stereotypical "caveman"—a Neanderthal or other pre–human hominid.[2]

To an archaeologist, context is essential. Before the invention of radiocarbon dating, archaeologists had no means to date artifacts other than via their context in the soil stratigraphy (layers) and/or surrounding artifacts. Even with scientific dating methods, sculpture and petroglyphs can't be dated scientifically because such tests would date the material from which the sculpture was crafted rather than the artwork itself. Some archaeologists have used this reasoning to reject the Dolní carving.

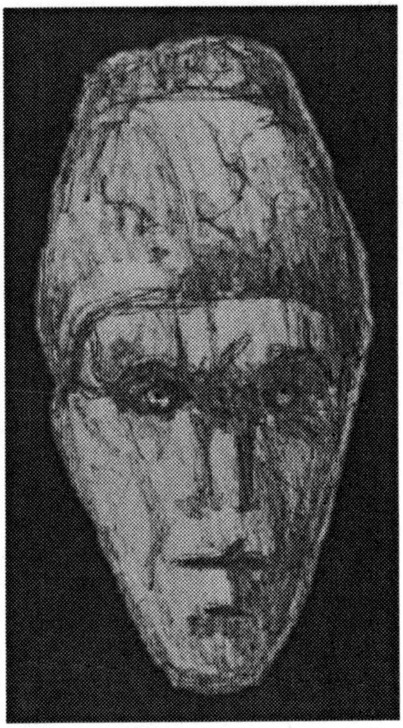

The female head found at the same location as the male head; both display a similar realism, especially in the eyes and noses (Dolní Vestonice, Czech Republic)—24,000 BC.

No one disputes that the ivory itself dates to about 26,000 years ago. Alpha–particle spectral analysis confirmed this date. Microscopic analysis both identified the tool used to carve the head, namely a flint implement, and revealed that sand and minerals had become embedded in the marks incised on the piece. Geologist Wakefield Dort, Jr. from the University of Kansas Space Technology Center, one of several scientists who examined the sculpture, believed the encrustation could not be faked. The formation of such a coating takes thousands of years. X–ray diffraction exposed the presence of iron oxide and fluorapatite, compounds known to stem from a lengthy burial.[41]

A female head also found at Dolní Vestonice and dated to 26,000 years ago displays a similar realism as the male head. As for the piece's sophistication, the cave paintings found throughout Europe show incredible sophistication. In fact, the earliest examples of cave painting are more sophisticated than the later ones. The sophistication evidenced by Paleolithic artists around the world outshines anything modern people achieved until the Renaissance.

Apparently, Paleolithic people didn't know they were supposed to follow a linear evolution in their artwork, or adhere to one style throughout their cultural lifetimes. They experimented, advanced, regressed, experimented more, and advanced again without concern for how their experimentation would confuse archaeologists 30,000 years later.

What the Dolní head most resembles is a hairy hominid. If creatures like America's Bigfoot have existed since the dawn of humanity, then we should expect to find representations of them in ancient artwork. Paleolithic humans may have considered hairy hominids as commonplace as bison, mammoths, and saber–tooth cats. Since they depicted those animals in their artwork, why shouldn't they have illustrated hairy hominids too?

If the Dolní head represented the sole evidence for ancient contact with hairy hominids, dismissing it would be easier. Yet further examples await us in the glaciated world of the Paleolithic.

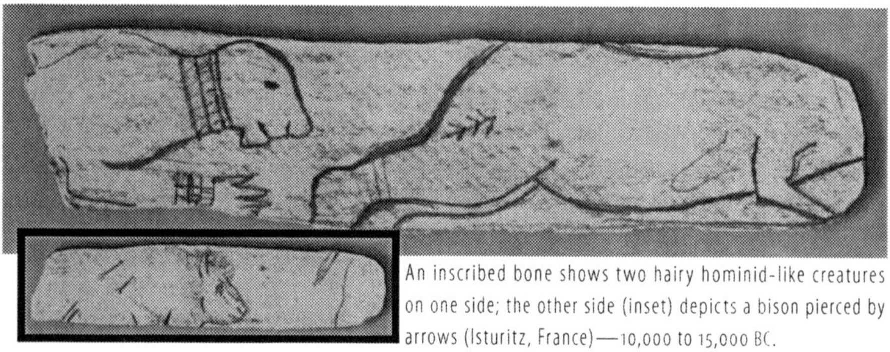

An inscribed bone shows two hairy hominid-like creatures on one side; the other side (inset) depicts a bison pierced by arrows (Isturitz, France)—10,000 to 15,000 BC.

The Venus Enigma

An inscribed bone from Isturitz, France, portrays two humanoid females lying supine. The females have many features that echo hairy hominids: sagging, pendulous breasts; stocky bodies; virtually no neck; wide shoulders; large buttocks; and stubby fingers. If they had heavy brows, the females would look like the sisters of the creature captured in the 1967 Patterson–Gimlin footage shot at Bluff Creek, California. The Isturitz bone dates from between 12,000 and 17,000 years ago.

Another example encompasses the many figurines found throughout Europe and the Mediterranean, from France to Malta and into Asia. The figurines are known as "Venus" figures because archaeologists presume ancient people intended them to represent "earth–mother goddesses." Once again, if no one knows what an artifact means, archaeologists assign it to the category of religious symbolism. But the Venus figurines bear an interesting resemblance to hairy hominids. The figures have large, pendulous breasts that sag down their chests toward their abdomens. They have wide hips, wide shoulders, little or no neck, and large thighs. Some even have heads that taper into cut–off cones. The first of these Venus figures come from the Upper Paleolithic, like the one carved onto a large block of stone at Laussel, France. The Laussel Venus is at least 20,000 years old.

The head of the Venus unearthed at Savignano, Italy, tapers into a cone shape, just as witness descriptions often say hairy hominid heads do, especially in North America. Archaeologists interpret the cone shape as a phallus symbol, yet why should we not consider that the artist intended to represent a cone–shaped head? A literal interpretation of the evidence seems warranted, as no one alive today can claim to know what ancient artisans thought when they crafted their artworks. The Savignano Venus also has the sagging, pendulous breasts and stocky figure familiar from Bigfoot sightings. The marble sculpture is about 22,000 years old.[12]

Far from Paleolithic Europe, in Siberia, a culture known as the Mal'ta Afontova fashioned their own Venus figurines. The Mal'ta Afontova settled near Lake Baikal, in the arctic regions of what is now Russia, about 22,000 years ago. Their Venus figurines bear a striking resemblance to the European Venuses, and many archaeologists consider the traditions related, despite the 5,000–mile separation between the cultures.[12] Like their counterparts in Europe, the Mal'ta Afontova Venuses call to mind the hairy hominids of today: no neck, stocky bodies, pendulous breasts.

The same artists who created the Venus figures also painted and sculpted realistic images of bison, mammoths, and other animals. So, to assume they based their Venus sculptures on real creatures does not seem fanciful.

After the Meltdown

By 10,000 years ago, the ice sheets retreated, signaling the start of a new age called the Neolithic or "New Stone Age." Not long after, the first civilizations arose around the world. Urban centers cropped up in Turkey and Israel, while cultures in the Andes of South America and in the fertile crescent of Mesopotamia began to flourish. In this new world, we find more signs that humans shared their environs with hairy bipeds.

The Dawn of Myth

One of the earliest flood myths comes down to us from the Sumerians, an early society living in Mesopotamia around 2000–3000 BC. The version that survives today comes down to us from the Assyrians and Babylonians who, like the Sumerians, recorded myths and information on clay tablets in a script called cuneiform, one of the earliest written languages, which resembles chicken scratches. On a series of such tablets, they recorded the Epic of Gilgamesh. The tablets relate legends starring a real king called Gilgamesh, who has many adventures, including his experiences during the great flood familiar from myths around the globe. Gilgamesh also befriended a hairy hominid named Enkidu. The tablets describe how the goddess Aruru tossed a hunk of clay into the woods, from which Enkidu formed. The hair–covered Enkidu wore skins, just as a few Bigfoot sightings speak of the creatures wearing skins. The tablets say Enkidu possessed great strength and stood twice as tall as the average man, two features prevalent in modern sightings.

Enkidu lives among the animals, drinking at their watering hole and eating grass in their pastures. When Enkidu destroys a trapper's snares, Gilgamesh suggests the trapper offer up a woman to Enkidu so that the wildman might renounce his kinship with animals. The ploy works. Enkidu forgets about the animals as soon as he sees the voluptuous woman—for six days and seven nights he mates with her. Could this be the earliest mention of interbreeding between humans and hairy hominids?

Jon Bodsworth / The Egypt Archive

A pottery vessel bearing an image of Bes, the hairy dwarf god who protected women and their babies during childbirth (Egypt)—14th Century BC.

Gilgamesh and Enkidu ultimately become friends, and adventures follow. When the duo kill the Bull of Heaven, the gods condemn Enkidu to death. The wildman dies after a long illness.

Around the same time that the Sumerians wrote about Gilgamesh, the ancient Egyptians began to formulate their pantheon of gods and goddesses. Among the Egyptian pantheon lurks a strange little figure known as Bes. Depicted as a hairy dwarf, and an ugly one at that, the god Bes was worshiped throughout the history of ancient Egypt. He served as god of marriage and gaiety, as well as protector of women during childbirth.[23] Statues and paintings show Bes with large feet, a stocky body, and no visible neck. A feathered headdress generally conceals his head. Occasionally he was represented as a bipedal lion, reinforcing the idea that he was a hairy being. Bes' wife, Beset, had similar features and also could be illustrated as a bipedal lion.[38]

Early myths from cultures around the globe speak of hairy, bipedal creatures—from the colossal to the miniature—who lived long before humans. In Australia, aborigines relate how hairy giants inhabited the continent in the Dreamtime, the era before human civilization, a time historians consider nothing more than myth. To the aborigines, however, the giants were quite real. According to the legends, 10-foot-tall hairy beings immigrated across the ocean from Indonesia. When they reached their newfound home, the creatures developed a taste for human flesh, forcing the natives to kill the giants in self-defense. Legends from other aboriginal tribes talk about similar creatures who predate humans, and who often enjoy the flavor of human flesh.[30] In the Americas, natives have told similar tales. From the human-size creatures said to roam the Cascade Mountains of the Pacific Northwest, to the Stonish Giants witnessed by the Iroquois,[53] hairy hominids have entered the folklore of many peoples.

In Mexico, since ancient times, the locals have described a class of hairy creatures known as the *ikals*. These three foot tall beings seem to fly—20th-Century stories say they use devices akin to rocket packs—and occasionally kidnap humans.[65] Other tales of hairy hominids kidnapping humans come from locales around the world. Several famous instances of kidnapping occurred in the early 20th Century in America, most notably the 1924 case of a British Columbia man named Albert Ostman, who said a family of hairy hominids held him captive until he managed to escape.

A series of global legends concerning the Pleiades constellation also involves giants. The stories, from cultures as diverse and far-flung as Australia's aborigines and the ancient Greeks, explain how the Pleiades formed when a giant chased seven sisters into the sky. The Greeks called their giant Orion, a name familiar to most people as the source of the name for the constellation. The Pleiades story told by the Kiowa Indians says a giant grizzly bear (hairy hominid?) chased the women into the sky.[57] Both the giants of the Pleiades myth and the Mexican *ikals* have also been associated with UFOs, a link I will examine in more detail later.

Celtic Contact

A 6th–Century BC carving discovered near a Celtic burial mound in Germany[21] demonstrates the possibility that the Celts encountered hairy hominids. The male figurine has wide shoulders, almost no neck, thick thighs, and enormous calves. The feet are missing. The head is cone–shaped. Archaeologists interpret the head's shape as indicative of a hat topping the head, mainly because of the line that divides the cone–shaped top from the rest of the head. But the line could also signify a heavy brow ridge.

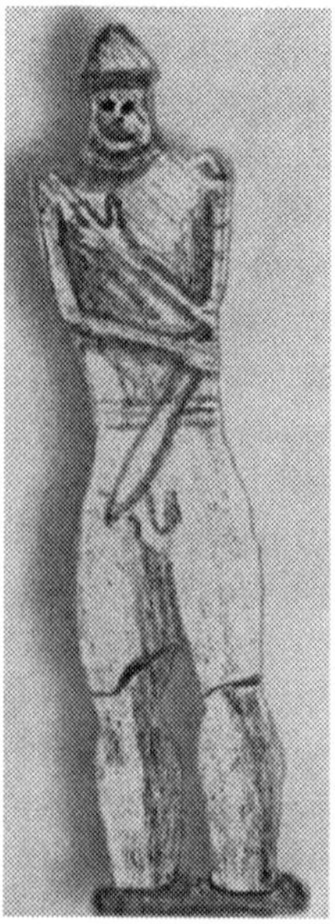
A Celtic statue with a cone-shaped head or hat (Baden-Württemberg, Germany)—late 6th Century BC.

Scottish fairy lore contains many references to the Brownies, pastoral spirits said to protect people they like. On the island of Inch, a Brownie watched over one family's cattle in exchange for a bit of milk the family would leave out for him. If they neglected to leave out the offering, the Brownie would kill one of their cows.[11] Similar stories of hairy hominids stealing milk from cows or taking food offerings left out for them come from modern–day residents of rural areas.

In the Scottish Highlands, tradition tells of a Brownie called the Glashan—a hairy man known for attacking women who ventured out alone. The woman would outsmart Glashan, though, gaining their freedom by hurling hot peat or boiling water at him. The Glashan would flee into the woods, howling in pain.[11] Witnesses who have shot at Bigfoot often say the creatures ran off howling or screaming.

A story from the island of Islay describes how two sisters, escaping from the third sister whom they had harassed, spent the night at the house of a giant. The giant ordered one of his minions to kill the human girls. In the darkness, the minion distinguished the human girls from the giant's daughters by the necklaces each wore—the human girls wore amber beads around their necks, while the giant's daughters sported necklaces made from twisted horse hair.[11] The story echoes the beliefs of South Americans, who are convinced hairy hominids braid the manes of horses.[22] The so–called myth of the Scottish giant also meshes with my own experiences, when my horses' manes have been braided mysteriously overnight (more on that later).

Through the Dark Ages and Beyond

Medieval Europe had the *wodewose*, the wild man of the woods, in whose honor the people held festivals. Fairy lore also provides many stories that echo modern sightings of hairy hominids. The *fions* of Brittany were hairy, swarthy dwarves similar in description to the *ikals* of Mexico. The *fions* were usually seen in the company of fairies, on the occasions when both would venture into the human world. The Secret Commonwealth written about by the Reverend Robert Kirk included not only fairies and leprechauns, but hairy dwarves of all sorts—as well as UFOs.[65]

Cynics might point out that the Dark Ages in Europe were a time of fear, persecution, and superstition. Myths of hairy hominids existed alongside stories about dragons, wizards, and demons. No one can dispute that the vilification of witches that ran rampant during the later 16th and 17th Centuries probably had its roots in the Dark Ages, when reason and knowledge fell victim to superstition. If the tales of hairy hominids sprang up during that period and vanished thereafter, I might agree. However, the stories do not cease with the lightening at the close of the Dark Ages. Instead, the evidence grows stronger the closer we get to the modern era.

Venus Returns

In the arctic regions of Eastern Canada and Greenland, a culture known as the Dorset emerged somewhere around 600 BC and flourished until about 900 AD. The Dorset culture produced figurines reminiscent of the Venus figures—and of modern hairy hominids. Archaeologists dismiss the figures as stylized humans, yet they share many features in common with hairy hominids. For instance, an ivory sculpture shows a female with sagging breasts, wide shoulders, no neck, and thick legs. Her hands and feet are chunky. Her face, though lacking a mouth, hints at a protruding brow. Few, if any, women—whether living today or in the ancient arctic—have the proportions evidenced by this statuette. One statuette, carved from wood, preserves the image of a humanoid with no neck. Its gender is difficult to determine,

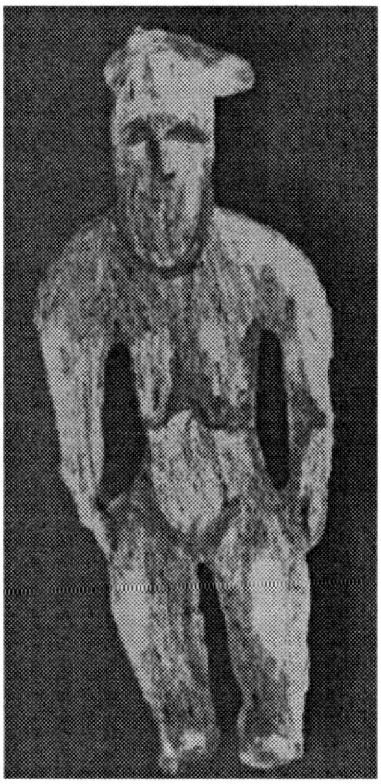

A Dorset figurine depicting a female with sagging breasts (Greenland)—600 to 900 AD.

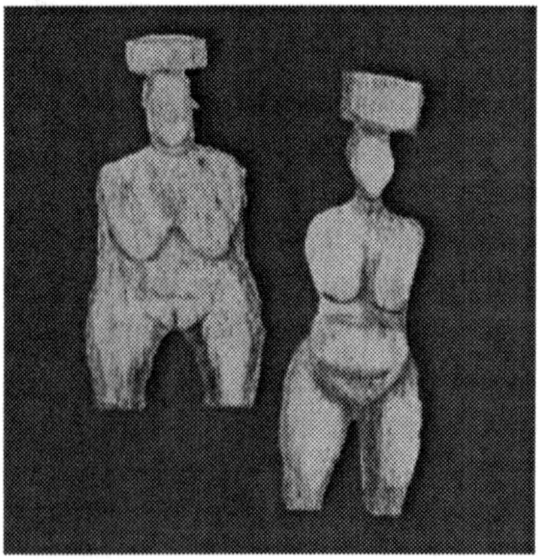

A figurine from the Thule culture, ancestors of the mdoern-day Inuit (Greenland)—1000 AD.

as the statuette has a hole in its chest and lacks both arms and legs. However, the humanoid does appear to have a heavy brow ridge and, possibly, a cone-shaped head.

In Alaska and Greenland, around 1000 AD, a culture branded the Thule surfaced. Considered to be the ancestors of the modern Inuit, the Thule culture produced figurines similar to the Venus and Dorset statues. A group of wooden figurines have sagging breasts, very wide shoulders and hips, thick legs, and little or no necks. Several even have tapered heads.[12]

Legends Reborn

Like many cultures around the world, the Eskimos include in their beliefs a creation story in which the first beings on the planet were giants,[36] as do the Norse myths and the Christian Bible. Norse mythology tells how the universe formed as a void, the northern portion of which consisted of eleven primeval, frozen rivers. The southern portion contained fire and lava. The fire in the south melted parts of the northern void. From this melting sprang two figures, the giant "frost ogre" Ymir and the cow Audhumla. Ymir then begat the other "frost ogres," the giants who first inhabited the Earth.[1] Ymir and the Frost Ogres also play a role in the Norse version of the global flood myth.

The Bible also tells of "giants in the Earth in those days" (Genesis 6:4), i.e. during the early days of mankind. The Biblical personalities encounter giants frequently, including the Anakims and the Emims, two races of giants: Numbers 13:33 says, "And there we saw the giants, the sons of Anak"; Deuteronomy 2:10 proclaims the Emims "a people great, and many, and tall, as the Anakims." Later, the Bible relates the story of the twins (fraternal, apparently) Esau and Jacob, the sons of Rebekah and Isaac. Esau is described as "red, all over like an hairy garment" (Genesis 25:25), while Jacob is human. The text also states that Esau belongs to a different race than his "brother." At one point, God informs Rebekah that "two nations are in thy womb, and two manner of people shall be separated from thy bowels" (Genesis 25:23). Years later, Jacob tells his mother, "Esau my brother is a hairy man, and I am a smooth man" (Genesis 27:11). The authors make certain we understand the difference between the two men—Esau possessed a hairy body and hailed from a distinct race of people.

Esau, the elder brother, is described as "a cunning hunter, a man of the field" (Genesis 25:27), while Jacob lives in tents. The hairy Esau also hunted and ate venison—deer, widely considered a staple of the hairy hominid diet. The Bible goes on to recount how Jacob stole Esau's birthright (his position as heir), and Esau later became ruler of the land of Edom. Chapter 36 of Genesis expends 43 verses enumerating Esau's descendants, who ruled the land of Edom. The story resonates with half-spoken truths, as if the metaphors disguise the true story of human beings supplanting the original inhabitants of Earth, the hairy hominids.

What if, like so many Bible stories, this tale serves as an allegory? Perhaps the Bible authors sought to explain the history of mankind and hairy hominids on Earth in a manner that wouldn't frighten their fellow humans, or perhaps they retold a legend passed down for generation upon generation, from a time so ancient that no one remembered the truth behind it. The story of Esau offers a haunting clue that perhaps hairy hominids and humans sprang from a similar source, with the hairy hominids emerging first, yet diverged into distinct species. As one of the most detailed legends of a hairy hominid, the tale of Esau deserves close scrutiny. However, like any legend, no one can say for certain how much truth remains in it.

I subscribe to the theory that most legends have a basis in reality, however distant and murky that origin may be. Before the invention of the printing press, written records were largely constrained to the upper classes. In ancient Egypt, for example, few people knew how to read and write. The hieroglyphic inscriptions in temples and tombs existed for the glory of the king and his elite, not for the average man. Even hieratic texts, which saw use in everyday contexts such as letter writing, remained a medium for the wealthy. For thousands of years, humans had one surefire method of spreading news—oral traditions. The stories in the Bible, the legends of the Norse and other cultures, must have stemmed from oral traditions.

Myths of giants and hairy beings do not end with the onset of the Christian era, however. In parallel myths from South America, both the Aztecs and the Inca reported that giants inhabited the world before humans. In the Inca version, the god Viracocha fashioned "stone giants," whom he then imbued with life. In the Aztec myth, the giants lived during an era called the First Sun, one of several cycles which the Earth endures.[34] The legends of Viracocha are echoed in numerous cultures throughout South and Central America. Everywhere people revered him as a civilizing god of tall stature with pale skin, blue eyes, and a beard.

Natives of the Solomon Islands also preserve legends of hairy hominids with whom they've shared the islands for thousands of years, extending into the present day.[6]

The Art of the Rocks

Rock art is notoriously difficult to date. Archaeologists often rely upon context in order to date artwork inscribed or painted on stone—this particular etching looks more primitive than another, therefore it must date to an earlier period than the more sophisticated example. Any discussion of rock art, therefore, includes speculation about the dates of creation. Scientific dating techniques, such as radiocarbon dating, generally require a sample of organic matter (plant fiber, bones, etc.). Though some rock art contains pigment, the types of paint used have proven as difficult to date as the etchings.

National Park Service

A Fremont petroglyph of a bipedal figure (Dinosaur National Monument, Utah/Colorado)—400 to 1450 AD.

Rock art encompasses three types of artwork: petroglyphs, created by scraping off the surface layer to expose the lighter rock underneath; picto-

Petroglyphs on Newspaper Rock (Canyonlands National Park, Utah)—as early as 1000 BC.

graphs, images painted onto the rock surface; and intaglios, or geoglyphs, which consist of images on the ground formed by moving surface rocks to expose the different-colored soil underneath. Cave paintings, like the famous images in the Lascaux and Chauvet caves in France, are pictographs. The Nazca lines represent a collection of geoglyphs.

Petroglyphs

In Australia's Ku-Ring-Gai Chase National Park, aborigines scratched out an unusual figure on the rocky ground. The figure looks like a man (it has no breasts) with no neck, large feet, and a stocky body. I've used those words so often in this book that I feel as if I should think of a different phrase to use! But I've used the words over and over because so much artwork throughout history depicts creatures who share the physical traits of modern hairy hominids. The artwork seems to show identical, or amazingly similar, creatures around the world and throughout time.

In Utah, a culture termed the Fremont flourished from 400 to 1450 AD.[48] The Fremont erected structures similar to the Anasazi's, as well as rock art with comparable themes. Like the Anasazi, they disappeared from the archaeological record. At Dry Fork Canyon in Utah, a Fremont artist pecked an image of a humanlike figure into the rock wall of a hillside. The huge figure belongs to a class dubbed "anthropomorphic figures." When archaeologists fail to find a "rational,"

National Park Service

Anthropomorphs in the Barrier Canyon style (Canyonlands National Park, Utah)—1000 BC to 500 AD.

normal explanation for a figure, they label it anthropomorphic, which means the figure resembles a human being yet also has features unlike any human.

At Dry Fork Canyon, one huge anthropomorph has wide shoulders, no neck, and a tapered waist. The arms and legs are incomplete, or have worn away from the sandblasting of the wind. A line across the figure's forehead seems to suggest a brow ridge. Comparable figures are also found at Dinosaur National Monument, a park which overlaps the northern end of the border between Colorado and Utah. The Fremont immortalized these figures in clay too, as demonstrated by a figurine found at San Rafael Swell, Utah.

At Newspaper Rock in Utah, several cultures inscribed petroglyphs on a single boulder. Both the Anasazi and Fremont, as well as the archaic cultures that preceded them, chose Newspaper Rock as a gallery for their artwork. Even the Navajo and Ute tribes left their marks on the rock.[48] The boulder is crowded with petroglyphs of animals, abstract symbols, and even Conquistadors! Alongside the usual fare, however, the ancient peoples etched anthropomorphic figures.

The Newspaper Rock anthropomorphs have shoulders wider than any human being. They have no necks too, and short arms and legs. Horns or antennae adorn their heads. While no sighting that I'm aware of mentions hairy hominids having horns, the appendages shown on the petroglyphs may represent the artist's attempt to ensure the viewers recognize the creatures as fearsome. The short arms and legs may be an artistic convention, or perhaps the local hairy hominids had shorter arms and legs than most. Another possibility is that the creatures seen by the artist (or artists) usually held their arms bent in front of them, giving the impression of shorter arms. If the creatures stood in tall grass, or knelt, they would appear to have shorter legs.

Another Southwestern culture, the Anasazi, thrived from 700 to 1300 AD.[48] Little is known about the Anasazi, as they left no written records and disappeared from the archaeological record before the arrival of Europeans. As they spread out across the Four Corners region where Colorado, Utah, Arizona, and New Mexico converge, the Anasazi left behind enigmatic ruins and rock art. Their cliff dwellings inspire awe. Most cling to the edges of high, steep precipices that are difficult and dangerous to climb. The Anasazi's rock art, while easier to reach than their cliff dwellings, has proven no less enigmatic.

The Anasazi created countless petroglyphs of anthropomorphic figures. Many have wide shoulders, little to no neck, bulky heads, and builds any sumo wrestler would envy. Some have tapered waists, but all have the broad-shouldered appearance of a Bigfoot. Some even have conical heads. Archaeologists attribute the conical shape to headdresses[18]; however, unless the Anasazi rise from their graves to explain their artwork to us, no one knows whether the conical shape stems from a headdress or an actual tapered head. Hands and feet, when depicted, are quite often large.[18]

A series of anthropomorphs near Butler Wash, along the San Juan River in Utah, resemble the Fremont figures mentioned earlier—except the Anasazi versions have no horns. Their shoulders are so broad that it defies logic to think the artists intended to depict humans. Similar figures can be found at sites across Anasazi territory.

Another type of petroglyph may illustrate the most famous part of the hairy hominid anatomy—their feet. Many petroglyph sites contain images of footprints. Some of the footprints have a concave side, like a human foot viewed from an angle. Others, though, have no concavity. In fact, the other prints are wider in relation to their length than human feet. Footprint petroglyphs appear in sites around the Southwest, from Three Rivers in New Mexico to Newspaper Rock in Utah. The Fremont, the Anasazi, and their predecessors the Mogollon[48]—who emerged around 3,000 years ago—scored these footprints on rock. Could they represent hairy hominid feet?

Hawaiian petroglyphs—which may date back as early as 200 AD[55]—illustrate strange creatures, humanoid types, with arms so long they extend below the knees. These figures have narrow waists like the Anasazi and Fremont figures, but also short necks and large, almost conical heads. Since some witnesses have described gangly hairy hominids with tapered waists, the Hawaiian petroglyphs could represent our old friends the hairy hominids.

Pictographs

The Anasazi also sketched pictographs that have survived the ages. At Canyon de Chelly, Arizona, we find painted figures of stout anthropomorphs with large feet and hands, long arms, and extremely wide shoulders. The feet appear quite large, and wider in proportion to their length than human feet.

A culture older than the Anasazi—known simply as the Archaic Period hunter-gatherers—created ghostly figures who also bear an uncanny resemblance to hairy hominids. While less clear-cut than most of the examples discussed thus far, the Archaic-Period pictographs, when added to the evidence pile, provide another ounce of proof. The Archaic period dates back at least 1,500 years, and possibly as far as 3,000 years.[48] The Archaic artwork is also referred to as the Barrier Canyon Style. In this style of pictograph, the anthropomorphic figures have the broad shoulders and long arms we've seen before—but they also have elongated bodies that create an other-worldly aura. Similar figures at the Archaic site of Sego Canyon, Utah, lack arms and feature huge eyes; another set of Archaic anthropomorphs resides at Canyonlands National Park in Utah.

Geoglyphs

Most people have heard of the Nazca lines, a series of geoglyphs in the Peruvian high desert. But geoglyphs appear around the world, from England to the US. In Chile, a culture known as the Arica created huge geoglyphs on hillsides in the northern regions, sometime between 1000 and 1400 AD.[59] One geoglyph, in the Lluta Valley, shows a humanoid with wide shoulders and hips, no neck, and a blocky head—a body type familiar to us from Bigfoot lore and the artwork discussed elsewhere in this chapter.

Artful Evidence

The evidence gleaned from ancient artwork and legends provides interesting clues; however, myths and drawings will never prove the existence of anything, especially not hairy hominids. I contend that proving hairy hominids exist should not be the focus of research into the phenomenon. Instead, we should concentrate on learning everything we can about these creatures, so that one day we might communicate with them on a meaningful level.

No amount of evidence will convince debunkers. Since most scientists fall into the category of debunkers—due to their obstinate denials of any and all evidence, based on scant knowledge and no real interest in the subject—we can expect no vindication from the scientific establishment. Seeking it has proven fruitless, exasperating, and a complete waste of energy. In the future, the task will become no less hopeless. Instead of concentrating on attaining the definitive "proof," we should analyze the data already gathered and strive to add to that store of knowledge.

We can learn a great deal by studying the evidence from sightings and the fossil record. Perhaps we can even identify these creatures.

2
Stick Signs

Humans have sign language for those who cannot hear or speak. We use physical markers such as traffic lights and no-trespassing signs to communicate the rules of everyday life. Why shouldn't hairy hominids use similar techniques to communicate with each other—or us—by proxy?

The answer is, they do. Sticks arranged in meaningful and distinctly unnatural displays serve as the medium for their nonverbal communication. I call these displays "stick signs."

Initiation

November 22, 2003.

The date holds no significance for most people, yet for me it would become the day my odyssey into stick-sign communication began. I lived in the woodlands of North Texas, west of Fort Worth, a landscape of rolling hills bounded on the west by the cliffs that marked the ancient shoreline of a long-vanished sea. The bones and fossilized footprints of dinosaurs have been found here. The trees grow shorter in Texas than up north, with oaks and cedars the most common. The soil consists of red clay topped by a sandy loam that takes on the consistency of talcum powder when dry, and mixes with the clay when wet to form a sticky red mud. Grass covers the soil, except where the gully-washers instigated by thunderstorms have eroded out miniature ravines, exposing the red clay.

The first stick sign I ever found , a simple cross formation (Texas)—November 22, 2003.

Behind my property, thousands of acres of woods stretched toward the horizon.

I had heard strange noises before. Something had also chased my horses, two of whom acquired mysterious cuts on their heads and chests. One night in the fall of 2001, I woke in the middle of the night for no particular reason. Unable to fall asleep again, I lay in bed listening to the night sounds. After a moment, I heard horse hooves thundering up the hill behind the house, coming closer with each beat of hooves. Within a few seconds, the horses had reached the barn area. When the horses feel peppy, they like to dash from the back field to the barn. On this occasion, though, one horse kept racing about even after reaching the barn. Curious, I knelt by the window at the foot of my bed, which overlooked the backyard and horse barn.

The moonless night offered no help. I could see the silhouette of one horse standing near the barn. The other horse, the one still galloping in circles around the barn and woodpile, must've been my black Tennessee Walker. He remained invisible in the darkness, but I could hear him gasping breaths out his nostrils in the manner horses do when frightened. The fence twanged, as if something had snagged on it.

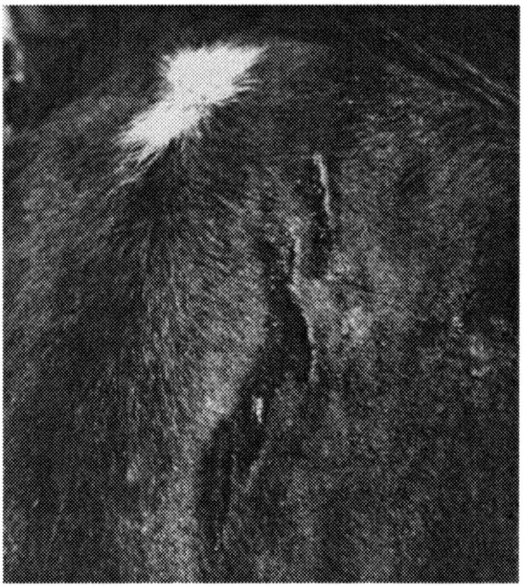

Cut marks my horse acquired overnight; they looked as if they had been scraped out by fingernails, but whose? (Texas)—July 22, 2003.

A man, even a tall one, must climb over the fence. Doing so leaves a telltale dip in the wire of the fence. Yet the intruder had left no dip in the fence, no shoeprints, no sign whatsoever of his passing. So who had chased my horse? I had no answer.

My interest in hairy hominids began years earlier, culminating in the writing of my novel *The Hunt for Bigfoot,* which would be published in January 2004. In November, however, no one had read my book save for a few friends and relatives. I had no inkling of what lay ahead for me.

Every morning I'd take my dogs for a walk around my property. The land was mostly wooded, with a three-acre field in the back and a clearing behind the house where the horse barn stood. About halfway between the house and the back field lay a manmade pond of the sort Texans call "tanks." Throughout the property, the horses had worn down paths through the woods, which the dogs and I would follow. On November 22, I set out on my usual morning walk. Along the northern path, beyond the pond at the eastern edge of the back field, I came across two sticks laid out in a cross formation. The sticks were short and stout. One stick rested neatly over the other to form a plus sign. The formation rested about ten feet from the path.

Although I photographed the formation, I dismissed it as natural. Sticks break and fall from trees. They get blown around by the fierce Texas wind. Sometimes they fall in formations that seem geometric.

Or so I thought.

The Stick Signs Return

Winter comes, even in Texas. The leaves desiccate like tiny mummies, littering the ground and masking the brown grass. Snowfalls are common in late December and January. Eventually, though, spring returns with a vengeance. Thunderstorms rip across the panhandle on a southeastward trajectory toward the Fort Worth-Dallas metroplex. Amid the chaos of spring in Texas, the stick signs made their comeback.

May 28, 2004. Once again, I'd set out with my dogs when a couple of sticks on the ground nabbed my attention. This time, a short stick sat propped up on a longer stick whose ends curved in opposite directions like a loose "S." Other sticks lay scattered on the ground nearby, but those were smaller and thinner and not laid out in obvious formations. I'd found the formation near the north fence line, along a path. It was just sticks, I told myself. Yet the discovery started a niggling in the back of my mind.

Two weeks later, on June 6, I found two more formations. The first, alongside the main trail that bisects the property, consisted of two chunky sticks: one short, one long, laid out parallel to one another. Two round fragments of bark, aligned between the sticks, resembled the crossbar of an "I."

The second formation I spotted along the north trail. Two large sticks lay parallel to each other, with a third stick curving out at an angle from below the first two.

Another cross sign (Texas)—June 6, 2004.

Now I began to suspect someone—or something—had placed the sticks on my property. I searched the area for footprints but found none. When I discussed the matter with other people, some laughed it off as my neighbors playing tricks on me. But my only neighbors were an elderly couple, a sheriff's deputy, and a fireman—hardly prime candidates for jokesters! No children lived in the neighborhood. The rural area was far from the nearest town.

Even if my elderly neighbor had climbed the fence in order to leave some sticks on my property, how could he have known I'd notice the sign? And what would've been the point?

Besides, I had six dogs who enjoyed free reign over the property and round-the-clock access to the outdoors via their dog door. Anyone trespassing on my land risked an encounter with six canine guards. Even the UPS man seemed wary of my dogs.

I had no idea what the stick signs meant. Yet I knew one thing for certain: my neighbors had not perpetrated a pointless joke. Who, then, had left the stick signs?

I continued to find stick signs throughout the summer, ten in June alone, mainly crosses and parallel sticks. One striking sign incorporated a bird feather, perhaps from a buzzard or crow. A long stick had been placed perpendicular to a much smaller stick, with the feather carefully aligned to the longer twig, its stem partly under the stick. The arrangement gave the impression the feather had grown out of the wood. The sign seemed like a friendly offering. But from whom?

Here's an interesting factoid, which may or may not have anything to do with the stick-sign phenomenon: in ancient Egypt, the feather was the symbol of truth and cosmic order. Ancient Egypt had its hairy hominids (remember Bes?) and—as I'll discuss later—also had its share of alien-looking beings. Is there a connection between Egypt, UFOs, and hairy hominids? I'll leave that for you to decide.

In this sign, a feather has been tucked partly under the stick (Texas)—June 28, 2004.

Artsy Interplay

I began to think of my night visitor as a "he," though I had no clue whether the being haunting my woods was male or female. The signs always appeared overnight. Moonlit or dark, cloudy or clear, cool or suffocated by humid heat—the weather failed to deter my visitor. He crept through my woods crafting artistic designs out of the materials at hand.

In July, the activity escalated. I discovered 38 stick signs that month. Again, parallel sticks and crosses dominated. The 4th of July ushered in a new episode in the saga, with an intriguing twist. A few days earlier, I'd decided to create my own stick sign in an area where I often found signs. By leaving my own sign, I'd let my visitor know I had caught onto the game. If humans were responsible, they might realize they could get into trouble for trespassing and stop their shenanigans. As my initial foray into sign making, I positioned four sticks into a square. As of July 3, my sign remained intact. By July 4, however, the visitor had rearranged my sign: two of the sticks now formed a "V," with a third placed as a crossbar atop the "V." The fourth stick lay parallel to the crossbar, a few inches away. Either I had extremely brazen, foolhardy grown men creeping around my property just about every night, or an intelligent creature on a par with humans had initiated contact.

I would've preferred to believe my neighbors had lost their mental balance and begun skulking around in the woods in search of sticks with which to tease me. At least then I could call the police. Unfortunately, more evidence suggested I had something quite nonhuman lurking in my woods.

A square sign I had made several days earlier was rearranged into this complicated formation (Texas)—July 4, 2004.

On the 8th of July, I stumbled upon an area of dirt that had been tamped down by some weight. The sandy loam all around the tamped-down spot looked its usual self—lumpy, strewn with leaves, and pockmarked with the imprints of horse hooves and dog paws, not to mention my own shoeprints. At first, I thought a small animal had curled up in the spot. But the area seemed an odd choice for a nap, due to its position smack on the northern trail through the property, a path frequented by the horses and dogs. The tamped-down area measured 17 inches long by 11 inches wide, tapering down to 8 inches wide at one end. A horse had stepped on the spot after its creation, leaving an imprint of the horse's frog (the interior of the hoof) and the leading edge of the hoof.

I took a photograph of the area. Then, to aid in my analysis of the tamped-down spot, I used a stick to draw an outline around it. The shape rang a bell in my mind. I'd seen similar shapes before.

During the extensive research I undertook for my novel, I read many books on Bigfoot and other hairy hominids. In several of those books, I'd seen photographs of impressions that resembled the tamped-down spot I had just discovered on my own property. In the books, the impressions were identified as the footprints of America's Bigfoot.

Had I stumbled upon the track of a hairy hominid? I needed more evidence before I'd be willing to say yes to that question—or ask the next logical question.

If a hairy hominid left its footprint on my property, had he also created the stick signs?

Dead possum, sliced in half with no blood, laid out 20 feet from the house (Texas)—July 7, 2004.

Strange Offerings

The day before I found the "footprint," one of my dogs had made a discovery of his own. In the backyard, about 20 feet from the house, my pit bull Rocky found a dead possum. Such a discovery, while gruesome, might appear unrelated to the stick signs and hairy hominids. The condition of the corpse, however, raised suspicions in my mind.

The body had been sliced in half in a manner so neat that it left no blood, no mess at all in fact. Some of the entrails were gone, but the body remained largely intact. Whatever had killed the possum had laid out the corpse with the hind end and head lined up, separated by a gap of perhaps six inches. The corpse rested by a tree.

My dogs had found dead animals on many occasions, and even killed a few themselves. Yet when one of the dogs kills an animal, that dog carries his prize around for all to see. When Rocky found the dead possum, he had no interest in disturbing it.

Thus far I'd found stick signs at the back of the property, away from the house. The day after the possum discovery, that changed.

The same day I found the "footprint," my visitor left a stick sign at the base of the tree by which Rocky had found the dead possum. The sign encompassed a large cross with two smaller, parallel sticks a few inches away.

Later that day, while exploring the back woods, I noticed a number of berries scattered over the ground right on the northern trail that bypasses the pond. The vines growing on the trees there bore fruit in the summer, dark-skinned berries the size of cherries. The vine from which the berries had fallen hung low, pulled down from the tree overnight. Horses don't eat the berries. Yet whatever had pulled down the branch had also scattered the berries, apparently, in its zeal to consume them.

I should point out that this region of Texas sports hard ground that even a posthole digger has difficulty penetrating. The talcum-powder consistency of the sandy loam, which swathes the hard clay, renders it useless as far as preserving footprints. Even horses' hooves leave nothing more than vague impressions unless a recent rainstorm has wet the ground. After a rain, the ground dries out swiftly. Shoes impress the best tracks, since the soles of most boots and tennis shoes have sharp lines on them. Bare feet generate almost no traces. If they should produce any traces, the shifting topsoil soon expunges the evidence.

The berry-eater had left no tracks. No human being would traipse around the woods barefoot, risking an encounter with the scorpions and rattlesnakes and stinging nettles.

Contact

On July 6, I recreated my square stick sign in a different location, this time on top of the hill behind the house. Many of the stick signs had

After I created a square sign, the next morning it had been altered into this shape, with one stick added on the left-hand side (Texas)—July 7, 2004.

appeared in this area. Overnight, my visitor altered my stick sign. He had picked up the entire sign in order to relocate it to a spot six inches east of where I'd constructed it. Then he had reassembled the four sticks into a square with one side angled outward, creating an opening like a door. Finally the visitor had added a fifth stick to the sign, a small twig sticking straight out from one side of the square. Once again, my visitor left no footprints.

I continued to find stick signs nearly every day. A relative joked that I should leave a stick sign in the shape of a tic-tac-toe board. "Why not?" I thought, and so I did create a tic-tac-toe sign on July 12. The next day, my nighttime visitor had rearranged the four sticks that comprised the sign into a series of parallel lines. More and more signs appeared as the days passed. I felt I was communicating with someone, though I had no idea what we were saying to each other. At this point I decided to keep going with the stick-sign exchanges, in hopes of learning more about the intelligence behind the responses.

On July 13, after seeing the changed sign, I created two more signs for my new friend—a square again, near where I'd placed the first square, along with the word "HI!" spelled out in sticks several feet from the square. About twenty feet east of these two signs, I crafted a third sign. This time I fashioned a triangle with, to amuse my visitor, a glow-in-the-dark dog toy inside it.

The following day, I received my reply. The visitor had transformed the square into a complex series of crosses. The sticks from the "HI!" sign had vanished.

As usual, no tracks gave away my visitor's presence. Over the next few days, I found more crosses and parallel sticks in the spots my visitor favored. The sign-maker started using one stick, snapped in half, to form the shapes. When I discovered these signs, the wood flesh would still be pale, signifying a recent break. The pieces always fit together, making it clear both had come from one stick. Unless sticks have a mysterious source of energy I don't know about, no natural force could pick up a broken stick then flop it down again in a "T" shape.

My visitor also began to incorporate non-stick materials into the signs—everything from blades of dead grass, reeds, and briars to remnants of fireworks my neighbors had shot off on the Fourth. The grass could've originated on my property, though not in the vicinity of the stick sign. As for the reeds and the fireworks, neither came from my property. The nearest crop of reeds grows a quarter mile down the road.

I spelled out the message "HI!" in sticks; the visitor first stole the sticks, then left this sign in the same spot a week later(Texas)—July 21, 2004.

One week after my visitor stole the sticks from the "HI!" sign I created, a new and complex sign appeared overnight in the same spot. The sign included an arrow alongside a number of sticks set both parallel with and at right angles to the arrow. On that day, my triangle sign had been changed too. The sign-maker had moved the dog toy from inside the triangle to a spot ten feet away. The triangle itself remained intact.

To top off that day, I found the first signs ever left in front of my house. Within 50 feet of the front porch, my visitor had crafted parallel rows of sticks.

Intruder Alert

July 22, 2005.

Another morning dawned in the weird woods of northern Texas, heralded by the discovery of an L-shaped stick sign 40 feet from my back porch, behind the horse barn. As I took the dogs out for our daily sojourn, I opted to travel the northern trail into the back woods. As the trail curved southward to meet the main path, the dogs caught a scent. They tracked the smell to a spot along the far side of the main trail, a few feet from the opening in the cross fence.

As I reached the dogs, I noticed that the ground they sniffed so furiously was soaked. In a patch about 18 inches square, the earth looked as if someone had poured a bucket of water onto it to drench the dirt below the surface. I shooed the dogs away to get a better look. What I saw puzzles me to this day. Someone had drawn a lopsided square—almost a trapezoid—around the wet spot.

The wet earth stank of urine. The dogs interest in it confirmed to me that none of them had urinated there and, besides, the soaked soil contained far more urine than a dog could deposit. The amount of urine also ruled out a human being. The biggest man couldn't let loose that much urine, even after knocking back a gallon of Gatorade. The urine smelled sharp and acrid like tomcat urine, only stronger—nothing like human urine. After rushing the dogs back to the house, I filled two sandwich bags with the wet soil. I compared the smell to that of horse urine, finding no match there either.

A lopsided square drawn around an area of urine-soaked earth (Texas)—July 22, 2004.

An L-shaped sign left on the same day as the urine sign shown on the previous page; as with all the stick signs, this one had no footprints associated with it, animal or otherwise (Texas)—July 22, 2004.

Strange offerings have appeared before and since the urine sign. The dead armadillo, rabbit, and possum all seemed like offerings of some kind. The hair shown below may serve as one too.

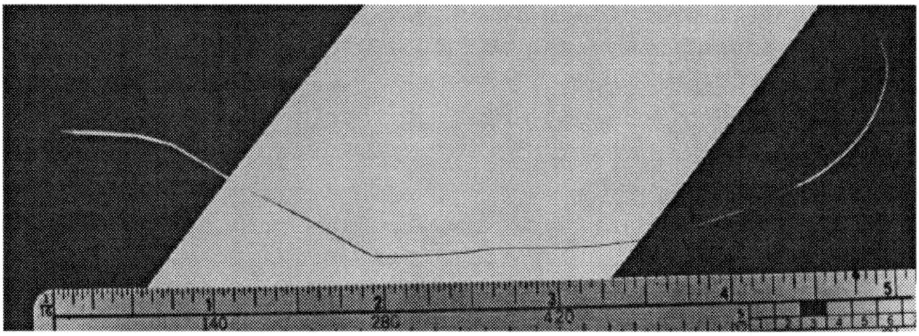

This long, coarse hair was found hanging on the electric tape fence that hems in the horse pasture behind the house. The hair does not resemble that of a deer or bear, it did not come from my horses, and it certainly doesn't belong to me. Some animals have long hair, but few have hair as coarse as this or colored like this specimen.

How the hair came to dangle from the fence remains a mystery as well, since electric tape fencing is neither sharp nor hard. The hair almost seemed to have been placed there on purpose. (Michigan)—June 8, 2005.

The lines drawn around the urine hinted at playfulness. My visitor had wanted me to notice the urine-soaked ground. Several reasons popped to mind: as an I-was-here statement; as a tactic to frighten me; or, as part of the creature's odd sense of humor. I may never know for certain.

If we try to understand a nonhuman creature's actions in the context of human behavior and thought patterns, we set ourselves up for failure. Hairy hominids may possess human-level intelligence but, if you examine the sightings, one fact becomes clear: they operate under mindsets vastly different from ours. You might call it a software incompatibility. Try loading a Windows program onto a Macintosh computer and see what happens. A similar situation may prevent us from understanding why hairy hominids behave as they do.

Cynics have suggested to me that no hairy hominid could create geometric shapes; therefore, one of my neighbors must've peed on my property as a sick joke. The fact that the lines drawn around the urine resemble a trapezoid may, however, be an accident. If you draw lines around something, those lines could end up looking like a geometric shape, regardless of your intent.

As for my neighbors, clearly they're not the culprits. Though I searched the vicinity for tracks, I found none. The amount of urine and the distinct odor associated with it indicate a nonhuman origin. I had hoped to have the urine DNA tested to prove its origin. Fate intervened, though. Shortly after finding the urine, I moved across the country to northern Michigan. I still have the urine in my freezer. Perhaps someday I'll find the means to have it tested.

In June 2005, I came across another strange sample. This time, the sample took the form of a hair of unknown origin that I discovered hanging on the electric rope fence around the horse pasture. The hair was black in the middle and white at both ends, long, and quite thick. The color and texture are wrong for deer or bear. However, the hair resembles samples found near the site where Bobby Clarke saw his Sasquatch. The local residents who found the hair also found large footprints, which led them to conclude the hairs belonged to a Bigfoot. We may never know for certain the origins of the hairs in their case or mine.

DNA tests may prove nothing anyway. Tests done on hair samples purported to come from Bigfoot turn out as "inconclusive." Without a type specimen—a body—on which to base our studies, we have no foundation for DNA analysis. Despite advances in science, the unknown remains unknown.

Escalation of Strangeness

The sign-maker refused to resign.

When I discover a new stick sign, I photograph it and then kick it apart so I won't forget I've already seen that one. Several times, however, the signs I kicked apart reappeared the next day, in the exact spot and in the exact arrangement as before. The interplay between myself and the visitor carried on in this manner throughout the summer. I'd create a sign, he'd change it. I'd wreck his signs, he'd recreate them.

The crosses and parallel sticks kept appearing. The signs grew larger. A cross would accompany parallel sticks, arrayed outward from the cross over the stretch of 10-12 feet. Signs appeared close to the house, as well as in the back woods.

Other physical traces emerged at the same time. My visitor seemed to have become either careless or accustomed to my presence. Perhaps our stick-sign exchanges convinced him I meant no harm. Whatever the cause, he left more evidence for me to ponder. On July 31, I noted scrape marks in the damp earth behind the house, near the tree line. While Texas has

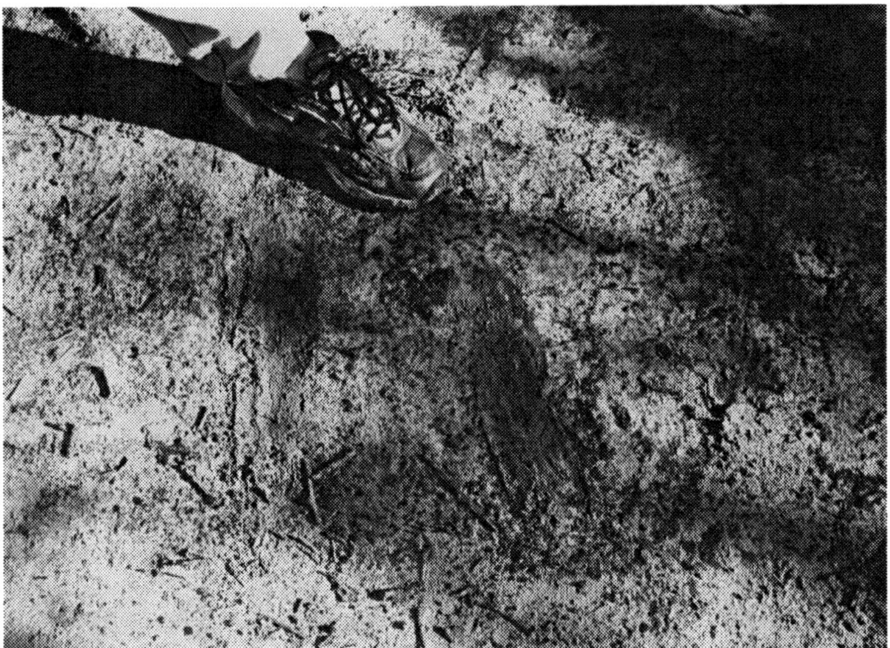

These scrape marks appeared overnight, just as the stick signs did; the dual marks almost seemed to end in toes or fingers (Texas)—July 31, 2004.

This large cross appeared in front of my house, 20 feet from the front porch (Texas)—July 31, 2004.

cougars, when cats or dogs paw the ground you can see the scrape marks from individual claws. The marks I saw that day looked like fingers or perhaps toes.

The same day, I discovered a large cross formation 20 feet from my front door.

Rising Tide

In August of 2004, activity skyrocketed. I found 64 signs on my property in a single month. Most of the signs consisted of crosses and/or parallel sticks; one included a curved stick about five feet in length, aligned with shorter sticks. Complex signs that combined crosses, "T" shapes, and parallel sticks also appeared that month. The most unusual, and striking, sign showed up on August 3.

As I walked out along the northern trail, the dogs alerted me to a dead armadillo lying on the ground at the entrance to the woods. The armadillo's hard skin had been sliced open atop and slightly behind the front haunches. Otherwise the animal seemed untouched. The innards had been removed. Whatever killed the armadillo likely ate the innards, since no remnants of them littered the ground.

No blood had spilled onto the dirt. I've seen dead armadillos before. My dogs have killed a few, but dogs leave behind a bloody mess. Whatever had killed this animal did so without leaving a mess, or else cleaned up the area afterward. The manner of killing the armadillo appeared almost surgical, as if the assailant had used a knife or razor-sharp teeth. The assailant must've overpowered the armadillo with ease.

Two years earlier, in November 2002, I had discovered a similar kill. Southeast of the pond, on the main trail between the pond and the back pasture, I'd happened upon a dead rabbit. Like the armadillo and the possum, the rabbit seemed to have been sliced in half. Back then, I blamed a cougar for the kill. Yet whatever killed the rabbit ate its entrails while leaving the rest out in the open. While cougars do eat portions of their kills and stash the rest, they generally move the carcass off to the side instead of leaving it out in the open for any scavenger to see.

Still, I might've dismissed the dead armadillo as the victim of a cougar or coyote too—until I noticed the stick sign. Within 18 inches of the carcass,

A dead armadillo, sliced in half with no blood or mess and the innards gone, was left on the north trail next to a stick sign; I had found similar kills on my property before, such as the partly skinned rabbit shown in the photo below (Texas)—August 3, 2004.

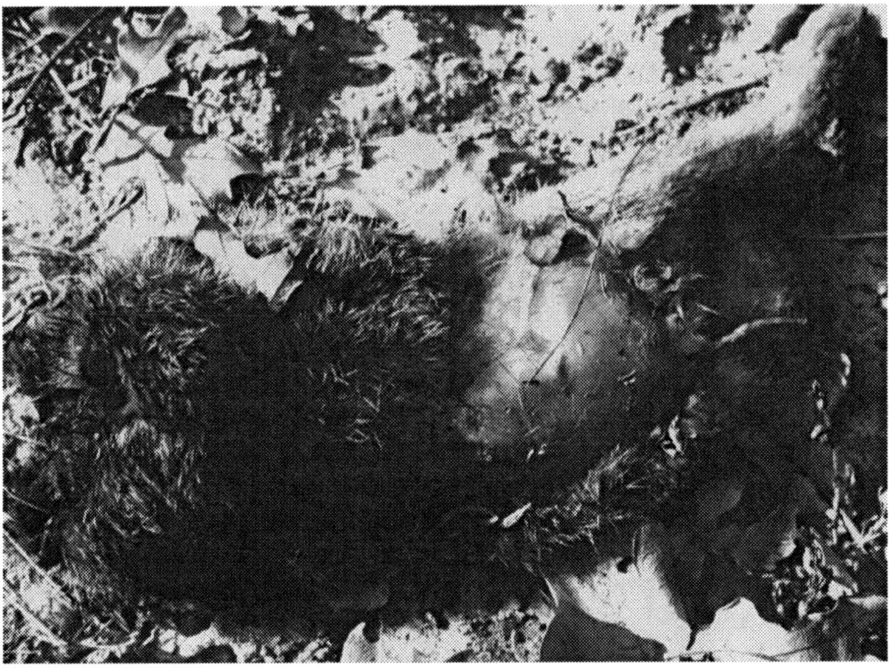

A dead rabbit found partially skinned and sliced in half, with no blood around or on it; this kill is nearly identical to the later one shown above (Texas)—November 7, 2002.

The stick sign that accompanied the dead armadillo; the cross was laid out two feet away from the armadillo's carcass (Texas)—August 3, 2004.

two sticks lay in a "T." Both the carcass and the stick sign sat right in the path, at the entrance to the woods, where I would be sure to see them. As usual, no footprints betrayed my visitor's presence. I began to wonder if my visitor intended the dead animals as offerings to me. If so, to what purpose? I may never know the answer.

Why no footprints? I often wondered this myself. With the exception of the vague print I found in July, I'd seen no other tracks associated with the stick signs or animal carcasses. The condition of the soil served as a partial explanation; the sandy loam simply refuses to hold a print. Yet I felt the answer must have a second part. Five days after the armadillo find, I came across a clue to the mystery of the missing tracks.

On August 8, on the north trail again, an odd pattern in the sandy loam caught my attention. After a rainstorm, the loam can look pockmarked. When an animal lays down on the sand, you can sometimes make out the impression of the hair preserved in the soil. The pattern I saw that day, however, looked unlike the patterns I'd become accustomed to seeing. In a strip roughly eight inches long by four inches wide, the sand was striated. Long, hair-thin, continuous bands swept across the ground as if something hairy had swiped over the area. For no reason I can quantify, an image popped into my head: the back of a hairy hand brushing across the soil.

Perhaps a hairy hand did create the impression. I compared the striations to those left when a dog lies on the ground. The striations left by the dog differed from those I'd found on the north trail.

An odd swirl mark that appeared on the same night as a stick sign (Texas)—August 8, 2004.

This "V" with additional sticks appeared on the same night as, and about 20 feet away from, the swirl mark shown in the photo above (Texas)—August 8, 2004.

Why would a hairy hominid brush his hand over the ground? If he'd glanced back to see his foot had left an impression on the ground, he might have quickly swept the back of his hand over the spot to obliterate the track. Few people would notice the striations in the soil. Few people would notice stick signs too, as I've realized since my first stick-sign discovery. Most people don't bother to look at the ground, except to search for footprints, and few would think twice about sticks lying on the ground. If your mind cannot accept the idea that an intelligent, humanlike creature lurks in the woods, you won't notice such things as stick signs and striated ground. Brushing away a track, after all, hints at analytical thinking. Stick signs indicate the same capacity for forethought and analysis.

On other occasions, I'd noted pockmarks in the soil when no rains had fallen beforehand. The pockmarks seemed larger and softer than marks hollowed out by rain. I wondered if the hairy hominid used more than one method to obliterate his tracks. To test my theory, I selected a branch that had fallen from an oak tree with its leaves intact and still green. I tapped the branch on the ground to erase my own footprints. The branch produced pockmarks identical to the ones I'd just found.

While researching *The Hunt for Bigfoot,* I became convinced that the evidence paints a picture of hairy hominids as humanlike creatures possessing an intelligence equal to ours, yet boasting greater physical strength and a physiology better suited to outdoor living. Hairy hominids could have a language too. The evidence led me to include in my novel the possibility that hairy hominids can recognize problems and deduce answers to them. I left a footprint by accident; oops, I'd better wipe that away in a manner least likely to attract attention. Using the back of a hairy hand to eradicate the evidence makes logical sense.

My nighttime visitor seemed to enjoy interacting with me, though from a distance. Yet leaving behind an obvious footprint would betray his presence and identity. Another possibility exists too—that I had multiple visitors. One enjoyed communicating with me, while the other preferred anonymity. To assume the playful creature had friends or a mate fits with the eyewitness accounts from elsewhere. Witnesses have seen family-like groups of hairy hominids. In a case from 1947, a couple in California watched as two hairy hominids crossed a highway in front of their car. The first creature, a male, helped his female companion traverse the roadway and clamber over a bank on the opposite side.[7] The sightings suggest hairy hominids share familial bonds. Therefore, we should expect to find groups (or pairs) of them sharing territory.

The Right Track

August 10. The Big Day.

I set out on my morning walk with the dogs, expecting I would find nothing more than a few stick signs. I did locate a stick sign on the north trail—several parallel sticks where none had been the day before, with a cross as the last element in the array. I followed the north trail as it curved southward to meet the main trail. The triangle sign I'd constructed by a tree remained intact, unchanged. When I turned onto the main trail, heading east toward the pond, something on the ground attracted my gaze—an impression 17 inches long by 9 inches wide. At one end the impression separated into four individual depressions, like stubby toes. I'd found a footprint.

I considered trying to make a plaster cast of the print. But the fragile, loose matrix of the sandy loam was unlikely to support the weight of plaster or maintain the impression if I tried to cast the print. I settled for photographing the footprint.

By comparing my photos of the track to photos in various books about Bigfoot and other hairy hominids, I determined that my track resembled footprints found in the Pacific Northwest, on sandbars and other areas with loose topsoil. The front of the print especially matched the photos other researchers had taken of hairy hominid tracks.

About 100 feet from the footprint, I spotted a second, partial imprint that showed the impressions of toes.

Five feet from the first footprint, I discovered another patch of striated soil. This time the patch measured about four inches wide by six inches long. The striations were clear, sharp, obvious. Either my visitor had left one footprint but wiped out another piece of evidence, or two creatures

A view of the entire footprint (shown in close-up on the next page) left by a hairy hominid, with a horse hoofprint alongside it; the footprint runs from just above the toes of my upper shoe to the toes of my lower shoe (Texas)—August 10, 2004.

A close-up of the toes in the footprint shown on the previous page (Texas)—August 10, 2004.

The second footprint I found 100 feet from the first (Texas)—August 10, 2004.

had strolled down the path—one erased her footprint while the other failed to notice he'd left an imprint. Hairy hominids, like humans, aren't perfect. Besides, the footprint was under a tree. The shadows cast by the branches could've masked the footprint, giving the creature the false notion that he'd left no tracks.

I discovered this patch of striated soil 5 feet from the full footprint; the marks appear to have been made by something hairy (the back of a hand?) brushing across the dirt (Texas)—August 10, 2004.

Sticks and Stones…and Acorns

After the second footprint find, the phenomenon underwent another transformation. Although parallel sticks and crosses abounded, other formations emerged as well. On August 12, I found a stick sign composed of four short sticks laid out in two parallel lines, as if they marked the corners of a square. The square layout accompanied a long line of parallel sticks placed a few feet away. The long line included sticks set down end-to-end along with a final grouping of parallel sticks arranged perpendicular to the end-to-end sticks. Near the east end of the long line rested a white stone. Neither the stone nor the stick signs had occupied that spot the previous day. No other, similar stones resided anywhere on my property.

The stone fit nicely in my hand, as if designed for handheld use. Even a larger hand would've had no trouble grasping the stone.

I decided to try a different tactic. Instead of leaving my own stick sign, I tucked a rolled-up magazine in the mesh of the cross-fence. Maybe my visitor would enjoy browsing the photos in *Archaeology Odyssey* magazine. At the least, I figured, a new tactic might trigger a different response. Alongside the magazine, I stuck two long sticks through the fence.

By the next morning, something had chewed on the sticks without disturbing them. The slightest nudge would cause the sticks to slide out onto the ground, but my visitor had managed to chew on the sticks either in place or by carefully replacing them after munching on the bark. My visitor hadn't finished after gnawing on the sticks, though. He also shifted the magazine two slots to the left. Two days later, I found the magazine on the ground 15 feet from the fence. One of the sticks lay on the ground too, while the other remained balanced inside the fence's mesh.

After that, I resolved to create a more complex sign of my own, to see how my visitor would react. A few feet off the main trail, I formed an arrow with a "T" attached to the point and a parallel stick adjoining the rear. Three days later, my visitor responded. The sticks I'd used had been removed. In their place stood one long stick with a V-shaped stick crossed over it a few inches from the end. The result looked astonishingly similar to the sign I'd devised.

The formations kept evolving as my visitor became bolder. Signs appeared three feet from the northwest corner of the house, right outside my bedroom window. Nothing more than a wall had separated me from my visitor as he crept past my window in the dead of night.

The realization of how close my visitor approached reminded me of an incident that had occurred several years earlier. At three in the morning, my dogs had run outside to confront an intruder. By the time their barking and snarling woke me, the fight had ended and the intruder had fled. I found the dogs huddled behind the garage. Two of them had gotten slashed. One had been cut on her belly. To acquire a slash on her belly, she would've had to jump up on the intruder. A cougar or coyote could not have sliced the dog in her belly.

I thought a human intruder had attempted to break into my house. But the intruder had left no footprints, no blood, no signs whatsoever. The dogs' wounds were clean and precise—like the wounds that killed the armadillo, the possum, and the rabbit. Had the stick-sign maker, who ventured so close to my window, gotten into a fight with my dogs several years earlier?

Sticks signs, stick signs, and more stick signs…as the days passed, I uncovered dozens of formations. My visitor fiddled with the sticks in the fence twice more—first, on the 24th, by removing them in order to set

The square arrangement that served as one end of a long line of sticks (Texas)—August 12, 2004.

This creek stone formed the opposite end of the long line (Texas)—August 12, 2004.

This response appeared several days after I had created a nearly identical sign; my visitor seemed to want to interact with me (Texas)—August 21, 2004.

them down in an X formation on the ground; the next day, after I'd replaced the sticks in the fence, by tipping them down at inverse angles to form a hanging X. I discovered the first triangle-shaped sign in August, along with another patch of striated sand.

I could've stayed up all night in the hopes of glimpsing my visitor. If I waited outside, though, my visitor would surely see, hear, or smell me no matter how carefully I hid. We are talking about a creature with human-level intellect supplemented by superhuman senses. By "superhuman" I don't mean superhero—I mean senses more acute than ours, perhaps more acute than any other animal's.

I suppose I could've waited up for my visitor inside the house, to watch out the window for his arrival. The security light installed by the utility company shone through the windows of the house, meaning anyone outside could see silhouettes of anyone inside. Besides, I would have no idea on which night my visitor would come. Unless I stayed up all night every night, I'd have little chance of picking the right evening. My visitor was too smart to show up on a regular schedule.

Other Bigfoot researchers have tried sundry methods of lying in wait for the creatures. None work. Hairy hominids let us see them when they please, not when it pleases us.

Finding a stick sign by the pool reminded me of a strange occurrence from two years earlier, one I had brushed off as nothing. The fence that delineates the southern property line approaches within five feet of the swimming pool. The security light skirts the property line in that same spot, while on the other side of the fence a stand of trees blocks the view. In May of 2002, I had noticed something unusual—just on the other side of the fence, right under the security light, some animal had laid down in the grass. A large area of tall grass and weeds had been flattened by whatever creature took a nap there. At the time, I figured deer had laid in the spot. However, the more I thought about it afterward, the less sense that made. The deer in Texas were shy, reclusive animals. They wouldn't lay down right under an extremely bright light, in plain view of a house and six dogs! I rarely saw deer, except in the back pasture, where they could flee into the woods upon sensing danger.

After I found the stick sign by the pool, I wondered whether an entirely different creature had taken a snooze on the other side of the fence.

Several times I set up my game camera by the swimming pool. My visitor refused to pose for a portrait. Two possibilities could explain why the game camera never captures my visitor. First, he could well observe me while I set up the camera and—though he has no concept of what a camera is—he realizes I've put up some kind of trap for him. Any creature who lives in the woods will have run across hunters' traps. The camera uses a flash too, which may deter my visitor. Unfortunately, night vision equipment demands more of an investment than I can make!

I freely admit the second possibility sounds rather outlandish, until you think about it a bit. The game camera consists of a normal, compact digital camera housed inside a camouflage-painted casing—a box equipped with a motion/heat sensor. The casing features a plastic window through which the camera snaps its images, along with a smaller portal for the sensor. The whole set-up gets chained onto a tree. I noticed how the camera window looks surprisingly like an eye. If I were a hairy hominid, and happened to spot the game camera, I might think the tree had an eye. I would certainly think twice about moseying past the tree to which the game camera clung. When the sensor triggered, the sensor flashes a red light that could be mistaken for eyeshine from an animal. We are talking about an intelligent creature here—not a banana slug. The hairy hominids need not understand the purpose

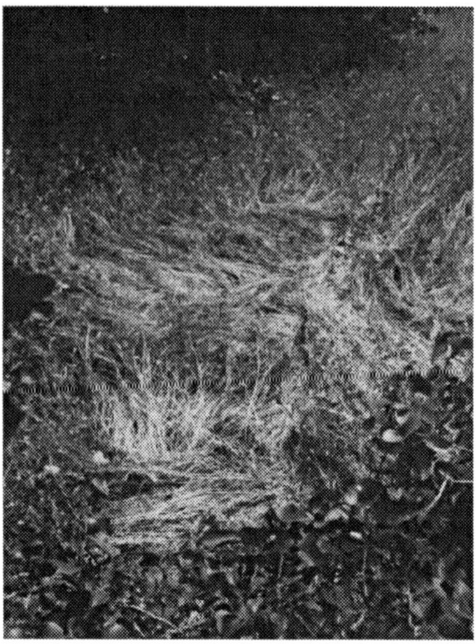

A grass bed made by some large animal, directly under a bright security light (Texas)—May 10, 2002.

of a camera, or its inner workings, to grasp the concept that the thingamajig on the tree has an "eye."

If these creatures are as smart as they seem, night-vision capability may not help. Other researchers have tried various kinds of video equipment, to no avail. If my visitor wishes to remain anonymous and unseen, he will.

In September, a total of 61 stick signs appeared on my property. My visitor also discarded stones on my property, types of rock that originate in a creek bed half a mile away. I'll have more to say about the stones in Chapter Four. As with all the stick signs, footprints, and other evidence, the stones appeared overnight in an area I hiked through every day. The stone invariably rested on undisturbed earth, ruling out the possibility a horse had kicked the rocks out of the ground.

Large, complex formations also emerged in September. Approximately 30 feet west of the house, a few yards from the driveway, I discovered an immense sign that encompassed five long sticks arrayed like steps—parallel to each other, but with each stick ahead of the last one. Alongside the sign, something had shaken hundreds of green acorns from the oak trees. I tried to shake the acorns from the trees myself, but they weren't ready to fall. The previous day, no acorns had fallen. Overnight, however, the immature acorns plummeted to the ground. To shake the acorns from the trees would require great strength.

A long array of stepped sticks found 30 feet from the house (Texas)—September 5, 2004.

Bones of Contention

On September 11, I walked out the back door with my dogs in tow, as I do every morning. Unlike other mornings, though, a strange offering greeted me in the backyard. Three feet off the porch, in front of a tree, lay a soup bone of the kind I give my dogs. While finding a dog bone in the backyard is hardly unusual, this one was different. A twig rested carefully propped up on the bone. I can't say for certain my visitor placed the bone and twig there. I know only that neither had sat in that spot the evening before.

The next day, I found an unusual sign in the back field. Small sticks had been laid out in a circle, with a four-inch gap in the perimeter. Two sticks jutted out from the gap, at right angles to the circle, like the neck of a bottle. The circle formation measured about two feet in diameter. In the coming weeks, my visitor would employ fragments of dog bones in other formations.

The last stick signs for September turned up on the 29th. Behind the horse barn, at the base of the hill, my visitor left behind a "V" with an additional, darker-colored stick next to it. Just over the hill, near the pond, he'd left another "V"—smack in the center of the main trail. Both signs seemed to point up the hill. What did they point toward? Perhaps nothing, for that's what I found in the area to which the arrows pointed.

While the visitor plainly intended to interact with me when he altered the signs I made, and when he left signs near the house, many of the other signs found in outlying areas of my property may have served another purpose. Since I find the majority of them along trails—paths frequented by deer and other animals, including me—I suspect most of them act as road signs. We erect signs to direct us on our paths, so why shouldn't hairy hominids do the same?

A circle of sticks with a hole in one side, like the neck of a bottle (Texas)—September 12, 2004.

October ushered in a new season, and 22 new signs. On the October 28, my visitor crafted three signs within five feet of the house. I would find no more signs before leaving Texas the next day.

Yooper Bigfoot

In September, I had decided to move way up north to the tip of Michigan's Upper

Peninsula. The Texas heat had never suited me particularly well. Some people love it, but I was not one of them. Besides, my family hails from Michigan, and I lived there as a child. For years I've wanted to go back. Finally, I just did it.

These two signs appeared on the same morning; the one in the above photo pointed uphill toward the one in the photo below (Texas)—September 29, 2004.

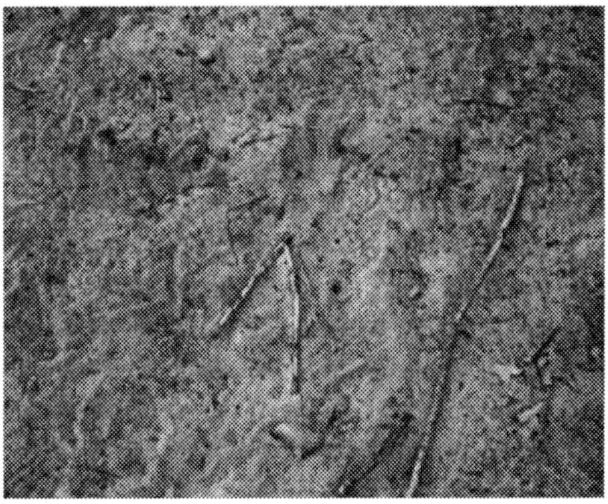

To fit in with my research into hairy hominids, I chose a remote area of the Upper Peninsula, known to Michiganders as the UP. Residents of the UP call themselves "Yoopers." My new home sits six hours from the Mackinac Bridge, a five-mile-long suspension span that connects the UP to the lower peninsula.

The Keweenaw Peninsula juts out from the western end of the main UP, toward Canada, surrounded on three sides by Lake Superior. Portage Lake and the Portage Ship Canal split the Keweenaw in two, with the Portage Lake Lift Bridge tying the halves together. I ended up here, on the virtual island north of Portage Lake. The big city up here, if you can call it that, is the college town of Houghton. Local souvenir stores sell t-shirts that echo how visitors feel about the region: the shirts depict a road sign that reads "Houghton—5 miles; End of Earth—6 miles."

My new home hunkers amid miles and miles of dark woods, much of it state land. Deer thrive here, despite the black bears, coyotes, wolves, and cougars. Hairy hominids seem to thrive here too. Everyone I talk to in the UP evinces a keen interest in the Bigfoot phenomenon. I've heard of sightings, though as yet I haven't been able to verify them. Because of the intense interest in the phenomenon, I established a research group called the Michigan Upper Peninsula Bigfoot Organization (MUPBO), a forum for people interested in hairy hominids, regardless of their backgrounds.

Everybody Has to Eat

The fact that the UP harbors a vast population of white-tailed deer plays into the possibility of a substantial hairy hominid population. Hairy hominids seem to prefer venison over other types of meat. In Texas, my night visitor favored the trails deer frequented, which also happened to serve as horse trails. While I'd found a dead rabbit, dead possum, and even a dead chicken on my property, I never found a dead deer. If the Texas hominids hunted deer, they would likely stash their kills in the dense woods off my property, rather than the cleared-out woods on my property. Since relocating to Michigan, I have found dead deer—or rather, my dogs have found them. The carcasses are hidden deep in the woods, where I rarely see them. My dogs sniff out the remains to bring back treasures of bone, hide, and sometimes hoof.

With wolves, coyotes, and probably cougars in the area, I can't say for sure what predator killed the deer. But the carcasses, secreted away in the woods, could well belong to a hairy hominid. When coyotes bring down their prey, they make a racket you can hear for some distance. Several times, I've heard the frenzied yipping of coyotes the night before my dogs unearth a fresh deer carcass. On other occasions, I heard no coyotes overnight, yet my dogs stumbled upon a fresh deer kill in the morning. The amount of deer scat I see in the woods could signify a population large enough to feed an army of hairy hominids.

The notion of Bigfoot as a gentle, vegetarian giant just doesn't mesh with the evidence. Nonhuman primates eat meat occasionally, though most of the protein in their diets stems from insects. Chimpanzees stand as the sole primate, other than humans, who hunt for and eat meat regularly.[60] However, in the arena of hunting, the chimps' skills are rather lacking.[49]

Sticks Up North

In Texas, I had no conclusive proof some crazed local hadn't made the stick signs and other physical evidence I found. If, after relocating 1200 miles away, I had found no more signs, cynics would have a valid point. Unfortunately for the doubters, my cross-country move failed to end the phenomenon. I had no neighbors on which to blame the signs anyhow.

In November 2004, my very first month back in the north woods, I discovered eight stick signs on my 40-acre property. One year had elapsed since I found my first-ever stick sign back in Texas. How fitting that my first Michigan sign should come in November too.

The signs I found that first month tended toward the complex. Several appeared on the dirt road that dead-ends at my property. One sign, which I spotted on the 10th, included a chunk of sandstone placed alongside a T formation. This region is famous for its sandstone, but the rocks don't grow on the roads! Another sign, left in the grass behind my house, consisted of five sticks arranged in a tic-tac-toe shape. Another, also left behind the house, employed 14 sticks to construct a square.

A "T" formation with a chunk of sandstone incorporated into it (Michigan) —November 10, 2004.

This square, comprised of 14 sticks, appeared overnight behind the house (Michigan)—November 15, 2005.

The first snowfall came later that month and, by mid December, the snow pack had swallowed the ground. Even the deer stayed in the deep woods.

In April, after the big melt, my new visitor left two signs. During the night of April 16/17, a sign appeared behind the house to the northeast. The sign contained two sticks shaped into a V that converged where the sign-maker had stood a third stick upright. I altered my visitor's sign by converting the V into a plus sign.

On April 23, I got an eerie hint as to the nature of my visitor. At 12:45 AM, I awoke because the bedroom had become stuffy. A window sits above my headboard, so I climbed onto my knees to unlatch and open the window. Just as I slid the window open, the brightest shooting star I've ever seen in my life streaked across the sky above the treetops. I knelt by the window for awhile hoping to see another shooting star. In the distance, a pack of coyotes yipped. I saw nothing else. But I heard something else …

Giving up on the meteors, I lay back down on the bed. The window I left open.

Off to the northwest, on my property, a creature whooped. The sound was loud, clear, and unmistakable. The vocalization sounded like "eee-YOOP"—a short call, high-pitched with deep undertones, as if it resonated from within a large chest. The call echoed off the trees, which means it had to originate within the five-acre clearing behind the house. A second later, overlapping voices answered from the distance. A moment later, a series of thumps issued from behind the house. I jumped up, certain someone was breaking down the basement door. When I looked out the west windows, however, I saw no one. I lay awake for over an hour listening for more calls. None came.

A "V" with another stick standing straight up out of the ground at the point of the V (Michigan)—April 17, 2005.

In an attempt to identify the noisy visitor, I ruled out the obvious suspects—coyotes, wolves, cougars, bobcats, bears, birds. Not even loons or screech owls emit such cries. When I heard the call, my first thought was "it's a person." But nobody lives nearby. The dense, dark woods and murky swamps hem in my property. Besides, I heard multiple voices that night. One weirdo who braved the swamps to hike for miles just to trespass on my property might sound plausible, but a whole troupe of weirdos?

I had heard recordings that were purported to be Bigfoot screams. When I could find no other explanation for the whoop call, I listened again to those recordings. The cry I heard on April 22 sounded similar to a whooping call recorded in 1974 in California, by the Berry-Morehead expedition.[67]

That night, the sign-maker altered my stick sign. He added two strips of bark, one parallel to the plus sign, the other at an angle, parallel to one leg of the plus sign. That morning, one of my dogs recovered a deer bone from somewhere in the woods. The bone had come from a recent kill. My dog hadn't killed the deer. Who had? Could the whooping calls and thumps play a part in the scenario?

Group hunting requires coordination. Perhaps a group of hairy hominids uses whoops and thumps to communicate during the hunt. The thumps could stem from hunks of wood (such as small trees) banged together, or from rocks struck against one another. Rocks would be easier to transport than large pieces of wood.

My visitor devised a new sign as well, which I also found on the 23rd. An old split-rail fence borders the road on one side, an incomplete boundary intended more as decoration than a barrier. I walk along this fence daily, and so I knew what it had looked like the day before—all rails in their slots, except the broken one, which hung down toward the rail beneath it. On the 23rd, another rail hung out of its slot.

What really drew my attention, though, was the rock sitting atop the fence post beside the dislocated rail. I had not placed the rock there. My dogs had not placed the rock there. My horses were innocent too and, unless ravens have developed an odd habit of plucking up rocks just to set them down on fence posts, no other animals deposited the rock there either. Since moving to Michigan, I also have no neighbors on which to blame strange occurrences. Even the strongest wind couldn't whisk away a rock only to plop it down on a wooden post.

Overnight a post in this fence (2nd from right) was popped out of its slot and a rock was placed atop the post (1st from right); the sandstone rock originated elsewhere on the property (Michigan)—April 23, 2005.

So who set the rock on the fence post? Maybe the same someone who whooped the night before.

May brought spring weather, along with 11 stick signs. Most showed up in the vicinity of the house. Why? Maybe my new visitors aim to welcome me to the area via stick signs, or perhaps they've always created stick signs around the house. This house was vacant for several years, so any hairy hominids in the region would've grown used to the house standing empty.

The rock that was sitting on the fence post, visible in the above photo (Michigan)—April 23, 2005.

The third possibility concerns the nature of the woods up here. Maybe I only notice the stick signs left out in the open. Twigs, branches, and uprooted trees litter the woods. The inept logging efforts of the previous owner inflicted much of the damage, while severe storms certainly added to the wreckage. If hairy hominids leave stick signs in the woods, those signs could well disappear into the debris. I have found signs along the game trails in the woods in the northeastern portion of the property, trails my horses have helped to clear of debris and weeds.

Like their Texas brethren, the Yooper hominids employed non-stick materials in their signs too, such as dead grass. They also snapped large sticks into pieces from which they formed signs.

The afternoon of the May 18, I heard a large animal crashing through the woods southwest of the house. I thought it strange that any animal would crash through there, as the woods in that area are free of debris. Still, I assumed a deer had made the racket when it fled from the dogs. The next day, I found a stick sign in the area where I'd heard the racket. The sign incorporated a cross, a third stick positioned at an angle to the cross, and a fourth stick jutting out of the ground. Exploring the vicinity, I discovered a possible footprint. It began to rain in the interim, turning the soft soil into mud and eliminating any possibility of casting the print. Photographs failed to preserve the track too, as the rain had already obscured its shape.

The 23rd brought the closest encounter yet, as my visitor created a stick sign on the front porch of my house, ten feet from the front door. Three days later, I found another set of parallel sticks, this time on the trail northeast of the house, near the garage. The two sticks that formed the sign looked as if they'd been broken, or bitten, off a larger stick. Each clean, precise break featured a notch in its center. Did the notches correspond to big, sharp teeth that had bitten through the wood?

Casting The First Stone

Stick signs have appeared throughout the summer of 2005. The hairy hominids in this part of the world seem to favor larger sticks posed in looser configurations. Some of the signs integrated bark strips into them—such as the sign composed of a stick abutting a circle of bark, the whole of which resembled a tennis racket. Another sign utilized a stick and a strip of green plastic mesh, of the sort used in doormats, to form a cross. I have no doormats. The mesh must've originated somewhere off my property.

Late in the summer, I began to hear banging noises emanating from the pole barn, a large metal building that stands about 100 feet from the house. The pole barn serves as both a garage and horse barn, with a side door allowing the horses entry into the building. At first I thought the horses had made the banging noises. But each time the noises erupted, when I looked out the window I would see the horses

This configuration appeared on my front porch, not ten feet from the front door (Michigan)—May 23, 2005.

back in their pasture, far from the barn. The noises start with one or two bangs followed by 10-20 seconds of silence, after which a volley of seven or eight bangs will finish off the cycle. I've heard the bangs at all times of day, in darkness and in daylight. By the time I get outside to see what's going on, the noises have ceased, though I do sometimes find rocks lying under the doors to the pole barn. Since the doors hang about six inches above the ground, the rocks could've rolled there after striking the doors.

Throwing rocks at various parts of the pole barn helped me figure out that the rock-thrower hurls his stones at the doors, which emit a different pitch when struck than the walls or roof. Pinecones tumbling from the trees above lack the hardness to create the sharp bangs I've heard. If I hurl a pinecone at the pole barn as hard as I can, the sound falls short of matching the pitch and volume of the mystery bangs.

Back in the horse pasture, a five-acre clearing behind the house, I've found rocks lying on top of green grass, in spots that hosted no rocks the day before. The horses didn't kick them up out of the ground—the rocks sit in undisturbed, grassy areas. One morning, on my daily sojourn with the dogs, I decided to head back to the house via the horse path through the woods southwest of the house. As I exited the woods, one of the horses spotted me. He jerked his head up, ears pricked, as he trotted a little closer in an effort to discern the nature of the figure stepping out of the trees. When I raised my hand to wave at him, he flinched. He remained agitated until I spoke to him, and he knew for certain it was me.

The horses' hooves have worn down the path to bare earth, so I knew nothing had sat in the path when I passed through at 10:00. Later that day, however, I discovered a rock lying in the middle of the path. I could not have missed the rock that morning. Situated directly in the path where I had walked, the hunk of granite stood out against the bare soil. Someone placed the rock in the path after my morning walk.

On July 30, I had observed a small piece of granite resting alongside the path that skirts the northern edge of the pasture. Nearby, a large chunk of granite lay in the grass. The large chunk, which weighed several pounds, measured about six inches across its diameter and bore scars where pieces had been flaked off it. The small, tapered stone measured about two inches long. When I compared the two stones, I noticed how the smaller piece fit onto the larger one right on one of the scars. The two had once been joined. The shape and size of the rocks, as well as the scars on the larger one, reminded me of artifacts I'd seen before—artifacts attributed to human ancestors. I'll have more to say about that in Chapter Four.

This sign found in front of the horse barn employed a small stick and a fragment of green mesh, like that found in doormats, to construct a "T"; the mesh came from nowhere on my property (Michigan)—August 3, 2005.

The Infinite Line

As I finish this book, in September 2005, the stick sign phenomenon persists. Like a line stretching out into infinity, the phenomenon shows no symptoms of ending. The stick signs act as a form of communication, both between one hairy hominid and another, and between the creatures and me—or any other human who bothers to notice the signs. Most people, in my experience, ignore what lies under their feet. We've been conditioned that sticks on the ground are sticks on the ground, nothing more, and that nonhuman animals lack the ability to communicate on a par with us.

Yet spoken words need not resemble human language in order to qualify as language. The Romans thought of the Germanic tribes of Northern Europe as barbarians incapable of speech. They rated the Germanic languages as equivalent to the baaing of sheep!

Language is relative. Our social perspective—and our biases—will influence what we view as language, culture, or intelligence. What we're willing to label nonverbal communication depends upon the same biases. If the method of communicating employs something other than symbols we recognize, from the Arabic alphabet to the hieroglyphs of ancient Egypt, we deny its existence.

What fate will the stick signs suffer? They may wind up ignored, derided, and denied, unless observant people take note of the earth under and around their shoes. You need no degrees or specialized training to spot a stick sign, photograph it, and document the find. But how can you tell a stick sign from twigs scattered by a windstorm?

First, you must explore and examine your surroundings daily. Take a walk with your dog, ride your horse, or hike with a friend. Notice what's there already—downed trees, fallen branches, loose twigs or stalks of dead grass—so you develop a picture of what the woods look like every day. Also take note of where certain plants grow, such as reeds or bamboo, and what game trails exist. Pay special attention to areas where you see deer scat (piles of small, oblong pellets). After inventorying your area, you'll have prepared yourself to notice anything out of place. The following list catalogs some key features of true stick signs:

- They appear overnight.
- They occur in areas where no other loose sticks can be seen lying around.
- Most stick signs are situated along pathways, especially those frequented by deer and/or horses.
- Parallel sticks and crossed sticks seem to dominate. Look for these occurring in areas that conform to the above guidelines.
- The best stick signs include geometric shapes, such as triangles or squares.
- Often, stick signs incorporate other media in addition to, or in place of, sticks. Watch for dead grass and reeds lying on the ground far from their source, rocks combined with sticks, or berries/seed pods scattered around the stick sign.

If you suspect you've found a stick sign, photograph it. Write down the location of the find, as well as a description of the circumstances in which you found it. Search for footprints or other sign, including mashed-down grass. You can install a game camera in the vicinity, if you like. Many researchers, myself included, have tried video surveillance with no success. In fact, the hairy hominids seem to avoid the cameras on purpose.

If you believe you've found stick signs, you can visit my website at www.upbigfoot.com to submit a sighting report. The more we learn about stick signs, the more we will understand about the sign-makers. We hold the power to save this unique brand of communication from oblivion.

A large cross that included a bent stick; like all the stick signs, this one appeared overnight along a trail frequented by me, my dogs, and my horses (Texas)—July 2, 2004.

A sign fashioned from one chunky stick split in half, with a small fragment nearby; the pale color of the flesh indicates a recent break (Texas)—July 20, 2004.

A set of parallel sticks (Texas)—July 21, 2004.

A blade of dead grass, not found in the surrounding area, delicately crosses over a T-shaped sign alongside a smaller stick (Texas)—July 21, 2004.

Here the dead grass & sticks sign in the previous photo has reappeared 3 days later, after I had destroyed it ; the sign was recreated using the same sticks (Texas)—July 24, 2004.

A group of Sticks laid out in a long, stepped line (Texas)—July 21, 2004.

A "U" sign that appeared near the fence on which sticks had been moved (Texas)—August 13, 2004.

These large, parallel sticks showed up behind the horse barn one night (Texas)—July 26, 2004.

I had balanced this stick in the fence; someone removed it, chewed on the end, and replaced the stick in the fence (Texas)—August 13, 2004.

Someone had messed with the sticks in the fence several times before this. A horse would not chew on sticks in this way and, besides, I have never known a horse to balance a stick in a fence! Whoever kept playing with the sticks had more manual dexterity than a horse. As for my neighbors, they couldn't gnaw a stick in this fashion.

Another cross, this one incorporating a chunk of tree bark; the sign-maker had begun to use materials other than simple sticks (Texas)—August 22, 2004.

The sticks I'd balanced in the fence had been tipped down ; doing so required great care, as the sticks would easily fall out (Texas)—August 22, 2004.

This complex sign involves four sticks, one of which is "Y" shaped (Texas)—August 24, 2004.

Once again, the night visitor had played around with the sticks in the fence; this time the perpetrator had removed them in order to lay them down on the ground in a lopsided "X" arrangement (Texas)—August 24, 2004.

For the first time, the sign-maker had constructed a triangle; a "T" accompanied it (Texas)—August 26, 2004.

Another triangle appeared 3 days after the first one, in a different spot; this time, the triangle sat in the exact spot where I'd made an arrow sign a few days earlier (Texas)—August 29, 2004.

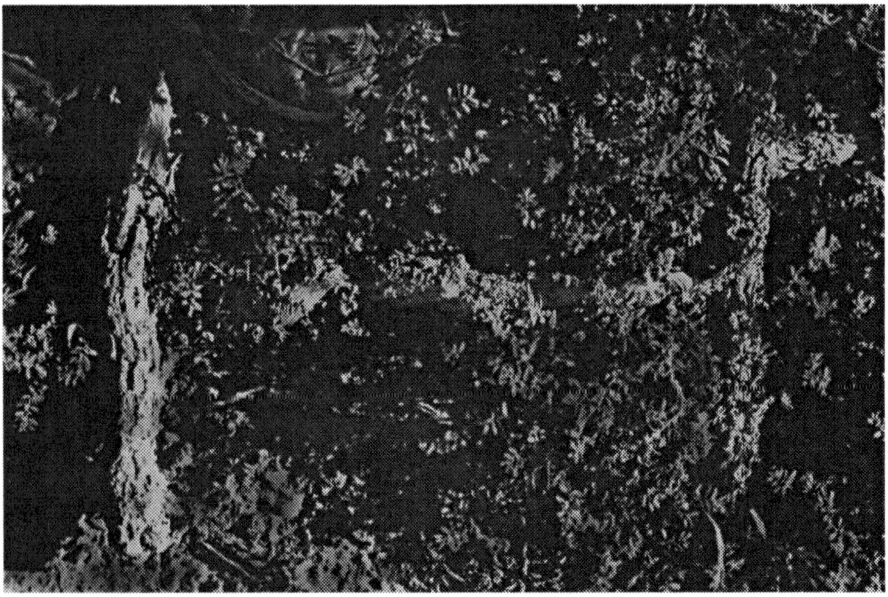

An "I" or "H" shaped sign, partly hidden in weeds (Texas)—August 31, 2004.

Some signs used materials not found on my property, like these pieces of reed (Texas)—September 1, 2004.

Note how the top reed has been bent into a curve without breaking it. The lower reed has been bent into a "V." The nearest reeds grew ¼ mile down the road. This sign appeared overnight, in the woods in front of my house.

I made the sign in the top photo on September 7, 2004. The next morning—September 8, 2004—the sign had been changed. No footprints, horse or otherwise, betrayed whoever had altered my sign.

Until this point, I hadn't thought about documenting the signs I made. I took pictures of the sign-maker's handiwork for my own records, and to show friends and relatives. By September, though, I'd begun to realize the importance of what was happening in my interactions with the sign-maker. From this point on, I would always photograph the signs I made for later comparison with the alterations my night visitor made to my signs.

A sign that showed up in my backyard, not 3 feet from the porch; the white stone is actually a fragment of an old dog bone that has been proppped up on the stick (Texas)—September 17, 2004.

Another sign that appeared within eyesight of the house; while the sign-maker could see the house from this spot, foliage shielded him from view as he made the sign (Texas)—October 9, 2004.

One of the first stick signs I found after moving north; this one appeared in the gravel road that dead-ends at my house (Michigan)—November 10, 2004.

This "X" crafted from strips of tree bark turned up in my driveway, a few feet from the concrete path to the house (Michigan)—May 5, 2005.

The stick sign in the left-hand photo was composed of one stick snapped into three segments; the right-hand photo illustrates how the pieces fit back together (Michigan)—May 11, 2005.

This sign appeared overnight in the woods southwest of the house; the center stick points straight up (Michigan)—May 19, 2005.

The stick sign in the above photo appeared overnight, as usual, on a trail northeast of the garage. The photos below show that the two sticks actually came from one stick, which was broken into at least two pieces. One piece displays an odd notch in its end. Could someone or something have bitten the stick to break it? (Michigan)—May 26, 2005.

A set of parallel sticks, widely spaced in the driveway near my car (Michigan)—June 5, 2005.

Top photo: During the summer of 2004, I'd begun making my own stick signs to see what response I'd get. I created this sign along a trail frequented by my horses (Michigan)—June 10, 2005.

Bottom photo: Several weeks later, the sign-maker gave me his response. It appeared in the same spot where I had made my sign, employing a stick and rock from my sign. (Michigan)—July 3, 2005.

Top photo: An "N" or "Z" shaped sign appeared at the intersection of two trails favored by deer; walking down either trail, you couldn't miss the sign. (Michigan)—July 23, 2005.

Bottom photo: This parallel sign appeared along a main trail frequented by the horses, northwest of the house. (Michigan)—August 1, 2005.

Above photo: The signs shown on the previous page were remade in the flatbed trailer parked a few feet outside the gate to my front yard, about 30 feet from the front door of the house (Michigan)—August 4, 2005.

Upon finding the two signs on July 23rd and August 1st, I had collected the sticks and taken them back to the house. There, I laid them down in a pile inside the flatbed trailer. The morning after I set the sticks from the parallel sign in the trailer, both signs had been reconstructed inside the trailer. While I had left all the sticks piled in a bundle, the sign-maker had known exactly which sticks he'd used before in each sign and reused the same ones to create his masterpieces again.

Below photo: This sign would be remade in the trailer too, after appearing next to my car one morning; strangely, the car's gas cap had been loosened (Michigan)—October 2, 2005.

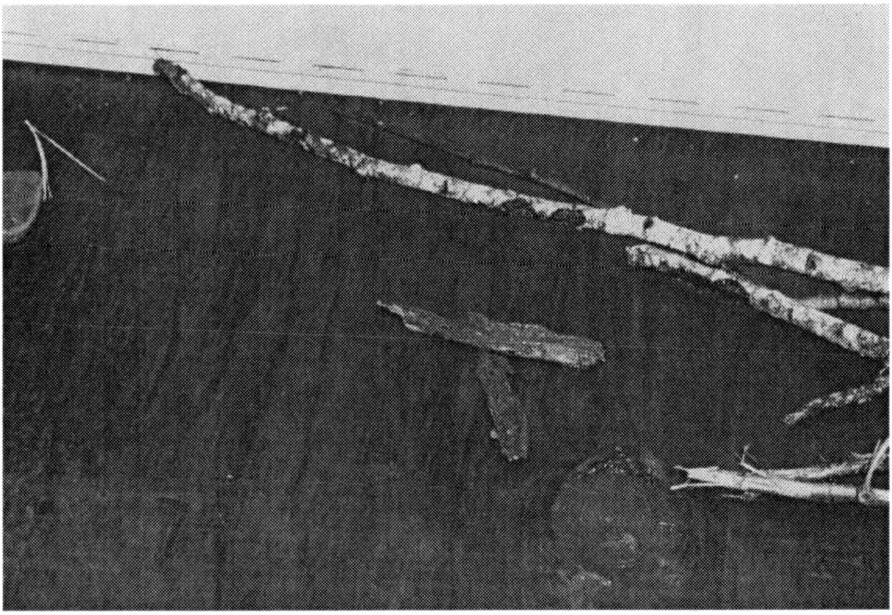

Top photo: The sign in the bottom photo on the previous page remade in the flatbed trailer; the sticks from the signs that had been remade earlier still lie in the trailer (Michigan)—October 4, 2005.

I had left the bark strips from the more recnet sign lying in the trailer, in a pile with the other sticks. The sign-maker recreated the sign just before a rain shower hit.

Bottom photo: A trio of parallel sticks that showed up in the horse pasture (Michigan)—November 4, 2005.

3
Dubious Origins

We've seen how hairy hominids have influenced the art and mythology of cultures dating back to the Paleolithic and beyond. The evidence suggests that our ancient ancestors both knew about hairy hominids and considered them important enough to warrant immortalizing them in art and legend. The Bible story of Esau hints at a common origin for humanity and hairy hominids, while simultaneously telling us each belongs to a distinct species. Stick signs demonstrate the creatures' intelligence, playfulness, and desire to communicate with us.

Before we try to identify the beast, however, we need to explore the murky origins of humanity, for only with a clear picture of our own past can we begin to comprehend the true nature of hairy hominids.

Darwin's Disciples

Schoolchildren learn that human beings evolved from apelike bipeds who in turn had evolved from the common ancestor shared by all primates, including humans. This forms the basis for the theory of evolution, the popularity of which we owe in large part to Charles Darwin. Others before him laid out the framework, and many after him filled in the walls, but Darwin remains the best known architect of evolutionary theory. At the same time that Darwin formulated his ideas about natural selection,

Alfred Russel Wallace conjectured the same theory. Yet, in later years, the men would diverge in their loyalty to evolution.

Over the years, the disciples of Darwin have come to call themselves paleoanthropologists. Like other anthropologists, they study humans and our relatives (chimpanzees, gorillas, etc.). But paleoanthropologists focus on the ancient past, interpreting the fossil record in order to understand how human life emerged and evolved. While archaeologists focus on the cultural remains left behind by humans, paleoanthropologists search for physical remains—bones—though oftentimes the two disciplines overlap.

Darwinian evolution consists of two principle tenets, common ancestry and natural selection. All species on Earth, Darwin believed, derived from one common ancestor, the trunk of the tree of life. From that trunk, species gradually branched out into new kingdoms, phyla, etc., until we arrived at the diversity of life we have today. Common ancestry implies another precept of Darwin's theory: descent with modification. Each generation of descendants of the common ancestor changed just a little from the previous generation, a process of gradual alteration that eventually led to new species. Natural selection stems from this process of change. Since some alterations are more successful than others, the successful changes will be perpetuated while less successful ones will gradually disappear as the offspring die before they can reproduce. Darwinian evolution requires that old species transmute into new ones, thus modifying themselves into extinction.[64]

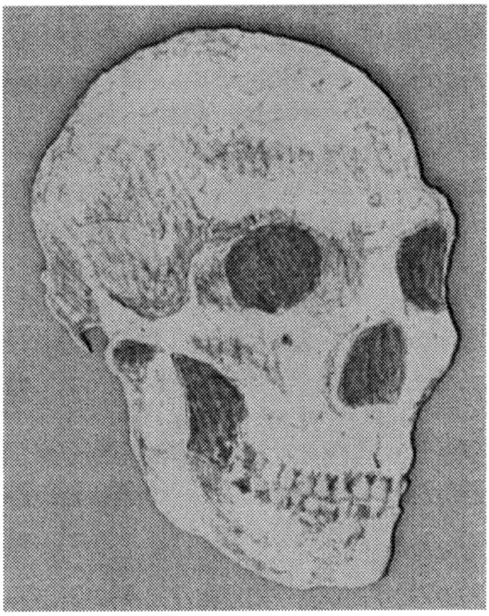

The skull of a *Homo erectus*.

But what is a species? Most people don't know the answer. The truth is, evolutionary science has no concrete definition of the term either. Most anthropologists would define a species as a discrete group of creatures who can reproduce amongst themselves and who resemble each other more than they resemble the members of other groups. They should not be able to interbreed with other groups. Yet, in reality, paleoanthropologists apply the term "species" to the fossil record rather arbitrarily. Nobody wants to become the next guy who discovers more remains of *Homo erectus*. Every paleoanthropologist dreams of

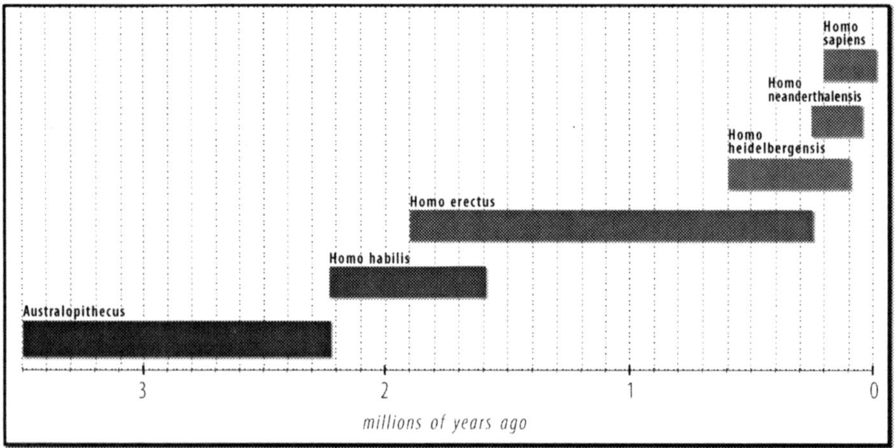

A typical chart showing the emergence of hominid species—*Australopithecus* includes several subspecies, the number and names of which vary depending upon the source; sometimes *Homo neanderthalensis* (the Neanderthals) is demoted to a subspecies of *Homo sapiens* known as *Homo sapiens neanderthalensis* and we humans become *Homo sapiens sapiens.*

unearthing a brand new species, a feat which reaps you fame, accolades from peers, and perhaps a cushy grant. So, when faced with a set of bones just collected from the badlands of Africa, what else can a paleoanthropologist do? Invent a new species, of course.

For the first century after Darwin presented his theory to the world, paleoanthropologists named so many new species of hominid that the family tree listed from the weight. Paleoanthropologists often declared their new species based on scant evidence: Davidson Black announced *Sinanthropus* based on three teeth found in a Chinese cave.

With paleoanthropologists in a fever over who would discover the next new "species," and willing to rally the slimmest evidence to become that person, the family tree seemed ready to topple. *Paranthropus, Atlanthropus mauritanicus, Homo modjokertensis, Homo primigenius, Sinanthropus, Telanthropus capensis,* not to mention enough varieties of *Australopithecus* to make your head ache...most of these species have since been relegated to the scientific dustbin. "Lumping" had replaced "splitting" as the order of the day.

Still, many paleoanthropologists of today harbor "splitting" tendencies, and the debate over what constitutes a new species rages as vociferously as ever. The debate over what constitutes a human has proven even more contentious, for species like *Australopithecus* display apelike characteristics in their anatomy, yet most paleoanthropologists consider that species a hominid—a human ancestor.

With most of the nomenclatural clutter cleaned up, paleoanthropologists may have hoped the worst lay behind them. Darwin's theories had melded with newer ideas to form a sort of unified theory of evolution, which dictated that modifications (known as mutations) caused ancestor species to transform gradually into new species, with natural selection as the engine driving the movement. Everything seemed in line, neat and orderly. Except…

The fossils refused to cooperate. If change occurs gradually, then the fossil record should illustrate this fact by yielding one fossil after another that slowly metamorphose into a new species. But the fossil records shows us the exact opposite—species pop up from nowhere, thrive for awhile, then poof! They're gone. The new species who appear magically in their wake may look similar to them, or they may turn out to be completely different. In 1972, Niles Eldredge and Stephen Jay Gould composed an article in which they pronounced that abrupt change, not gradual modification, fit the evidence. They dubbed this process "punctuated equilibria."

Yet they didn't dream up the theory without precedent. The early 20th Century saw the introduction of the Hardy-Weinberg principle, which employed mathematics to prove that populations of organisms prefer to remain static. Further research demonstrated the validity of the precept. The genes of interbreeding populations stay constant, when evolution requires that they must change. In fact, evolution requires that anatomical change, caused by the changing genes, produce new species. Yet this does not happen. Dog breeders, cat breeders, horse breeders—all can tell you that anatomical differences produced by interbreeding do not produce new species, a fact science still cannot deal with today.[64]

But, hey, the fossils still tell us a lot, right? Enough that we can form an accurate picture of human origins. Despite the arguments over which species such-and-such fossil belongs to, the outline of human ancestry has become clear.

Well…not quite.

Bits and Pieces

When Darwin wrote *On The Origin of Species*, hominid fossils were rare. Actually, fossils of any kind were rare. In the century after Darwin wrote his seminal work, the first paleoanthropologists fleshed

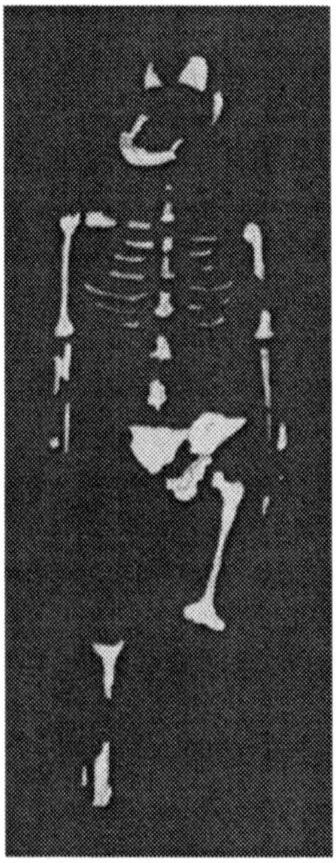

The skeleton of the famous "Lucy," an *Australopithecus*; she has retained just 40% of her shattered bones (Africa)—3,000,000 BC.

out the fossil record at an agonizing pace yet, slowly, the bones began to pile up in museums. Even today, though, the number of hominid fossils available for study remains quite small—and decidedly fragmentary.

Paleoanthropologists rarely dig for fossils. They scour the surface looking for bones that have eroded out of their burial spots. Every hominid fossil is found in fragments or, most often, shattered to bits by the geologic pressures exerted on them over the eons. The remains of the famous "Lucy," a three-million-year-old *Australopithecus* found in the 1970s, lay in shards scattered over a wide swathe of ground. Her skeleton remains only 40% complete: she has lost her hands, feet, half her pelvis, the majority of her legs, and most of her skull. Yet, in the wake of her discovery, paleoanthropologists waxed about the form and function of her body.[64]

Did her hands resemble an ape's or a human's? Did she use them to hold stone tools, or to swing from tree to tree? Did she walk fully upright, or stooped over? The debaters supported their ideas by describing the anatomy of her hands and feet—*appendages not represented in her remains.*

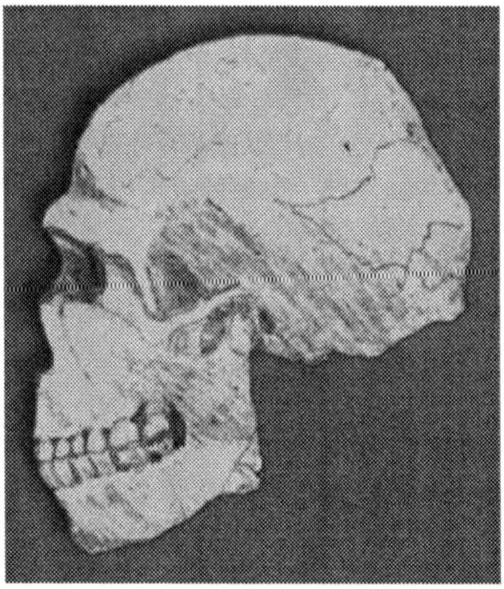

One of the skulls found in the cave at Skhūl, Israel, a site which has incited much debate.

Cave of Secrets

No paleoanthropologist worthy of the title lets a little thing like facts get in the way of a good theory. In 1932, excavators working at a cave site called Mugharet es-Skhūl, on the Mediterranean coast of Israel, found the remains of ten hominids. Despite the distinct brow ridges present on each skull, paleoanthropologists label the remains as ancient examples of modern humans.[64]

When outlining the differences between modern humans and ancient hominids, however, brow ridges are invariably listed as a key distinction. However, several factors point to the remains being Neanderthal rather than "early modern" human:

1. The stone tools found at the site come from a class known as Levalloisian-Mousterian, a toolkit attributed to—and found with the remains of—Neanderthals.
2. At the time of their discovery, the Skhūl hominids were classified as Neanderthals.
3. Other cave sites in the same area contained undisputed Neanderthal remains dated to between 120,000 and 60,000 years ago, while the Skhūl fossils date to 80,000[12]-100,000[62] years ago—making them roughly contemporaneous.
4. In Israel, paleoanthropologists can find *no meaningful difference* between the tools used by the Neanderthals and those used by "early modern" humans.[63]

On top of all that, the Skhūl remains resemble Neanderthals far more than modern humans. The Skhūl "people" sported sloping foreheads, brow ridges, projecting jaws, receding chins, and long braincases. Yet in their quest to prove evolution, paleoanthropologists need a "transitional" species between *Homo erectus* and modern humans. Against the evidence, they marshaled the Skhūl hominids to fill in the ranks of human evolution.

Hard as the paleoanthropologists struggle to reconstruct the landscape of human evolution, pitfalls trip them up at every turn. In order to have evolved into us, the ancient hominids had to improve their looks and brains with each new species. They had to demonstrate enlargement in stature, from the 3-foot-7 Lucy all the way to us, as well as enlargement in brain size, from a puny 450-500 cc in *Australopithecus afarensis* to the average of 1,040-1,595 cc into which 90% of humans fall[12]. The discrepancy in body size between males and females—known as sexual dimorphism—had in *afarensis* gone to extremes, with males twice as big as females. That disparity had to go too. The species that evolved from *afarensis* must lose such features in order to satisfy evolution.

Enter *Homo habilis*. In 1986, paleoanthropologists working at Olduvai Gorge discovered a partial skeleton that included, for the first time, arm and leg bones belonging to one individual along with portions of the skull. The excavators identified the remains as *Homo habilis*. Until then, few remains of the species had turned up; however, paleoanthropologists used what little they had to paint a picture of a hominid more robust, taller, and bigger brained than *afarensis*, for they needed such a species in order to fill in the gap-ridden image of human evolution. But the new specimen from Olduvai refused to conform.

The fossil, christened OH 62, shares more in common with *afarensis* than with its supposed descendant, *Homo erectus*. The apelike body, small brain, and sexual dimorphism seen is *afarensis* continued unchanged in *habilis*. Comparing the leg bones highlights the difference between us and them: in humans, the humerus is 70% of the femur's length; in *afarensis* and OH 62, the humerus is nearly 95% the length of the femur. The figure for *afarensis* and *habilis* is comparable to that of apes. To lose its primitive traits in the transition to *erectus* means that *habilis* had only 200,000 years in which to metamorphose.[40] Unless the entire population of *habilis* on the planet cocooned themselves, emerging six months later as *erectus*, the fossil record leaves us with a mystery. By the wave of an evolutionary wand, *Homo habilis* transmuted into *Homo erectus* virtually overnight. This little problem has led some paleoanthropologists to engage in wishful thinking, some by demoting *habilis* from the genus *Homo* to the genus *Australopithecus*[42], still others by splitting *Homo habilis* into large and small breeds[63]. What's next, cocker spaniel *habilis*?

Top it all off, recent studies have even cast doubt on the bipedal status of *Australopithecus* and, by way of their close association, of *Homo habilis* as well. Anthropologists at George Washington University scrutinized the wrist bones from several examples of *Australopithecus*. They concluded that the species had wrists designed to lock, suggesting *Australopithecus* would've preferred knuckle-walking (a quadrupedal gait) to walking on two feet.[9] The living apes prefer knuckle-walking too, or swinging from tree limbs, which indicates the early hominids had a stronger affinity to apes than to modern humans.

Island Hopping

The hominid tree has only gotten more crowded, and the branches more tangled, in recent years. In the mid 20th Century, priest and amateur archaeologist Theodor Verhoeven discovered stone tools on the Indonesian island of Flores. Archaeologists now know the tools date to about 840,000 years ago, back when no hominids should've lived on Flores. The island has never been connected to the other islands in the archipelago, or to mainland Asia. The tools forced archaeologists to consider the possibility that *Homo erectus* had made the oceanic journey nearly one million years ago. But no *erectus* remains have been found on the island. As the debate over the stone tools continued, a new find flipped conventional wisdom on its head.

In 2004, paleoanthropologists excavating in a cave on Flores, expecting to find some *erectus* bones, instead unearthed the remains of a startling new hominid—a child-size creature with a brain smaller than a chimp's. At first, the archaeologists thought they'd dug up the skeleton of a modern human child. But closer inspection proved the bones belonged to an adult, though an adult what they still didn't know. They dismissed the possibility of a modern human deformed by disease or starvation.

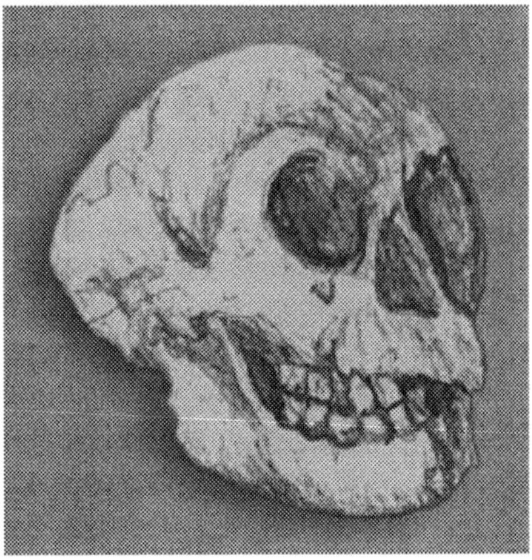

The skull of *Homo floresiensis*, a diminutive creature recently added to the family tree of hominids; the species survived until about 12,000 years ago.

Soon, the archaeologists realized they'd uncovered a species of dwarf hominid, which they promptly christened *Homo floresiensis* and nicknamed the Hobbit. The Hobbit represented a heretofore unknown species that had alighted on Flores at least 95,000 years ago and survived until about 13,000 years ago.[46] Since modern humans arrived on Flores about 50,000 years ago, the first humans must've shared the island with the dwarf hominids. Yet dogma states that humans coexisted with no hominid species that late in our history. Once the Neanderthals died out, we were alone.

The Australian-led team of archaeologists has since found the remains of six other Hobbits, including a portion of a second skull, along with stone tools of finer quality than the tools Verhoeven had discovered half a century earlier.[46] The Hobbits, despite their petite brains, appear to have sported intellects more advanced than the bigger-brained *Homo erectus*. Given this discovery, plus the lack of *Homo erectus* fossils on Flores, archaeologists and paleoanthropologists have had no choice but to consider the option that *floresiensis*, with its teeny brain, fashioned all the tools found on the island.

Other paleoanthropologists have tried to discredit the Flores find. The remains obviously belong to an ancient ancestor of the pygmies who inhabit the island today. Yet pygmies, while small-bodied, have large brains[68]—the Flores hominids have miniature brains to match their elfin bodies. Next, critics rallied behind the concept that the original skeleton represents an ancient, but modern, human afflicted with a rare disease called microcephaly. Individuals with microcephaly have shrunken, deformed brains and dwarfish bodies. Yet analysis of the Flores skull revealed no deformation. The Hobbit's brain is simply tiny. The other bones found at the site, as well as the partial second skull, prove the find was no anomaly. Rather, the original skeleton stands as the type specimen for a new species of hominid.

The Flores hominids seem like a hodgepodge of all the hominids that came before them. Their skulls resemble those of the smaller, earlier *Homo erectus* found in Africa. The pinching of the skull at the temples suggests a kinship with the hominids from Dmanisi, Georgia, a species whose identity paleoanthropologists still debate. In still other ways, the Floresians resemble *Australopithecus*.[46] Another fact adds to the mystery—the last hominids this petite, members of the species *Australopithecus*, lived three million years ago *in Africa*.

So from whence did *floresiensis* come? The researchers who announced the species have explained the Hobbits as *Homo erectus* who became marooned on Flores by some unexplained means, then succumbed to "evolutionary pressure" which forced them to shrink in size (and, apparently, brains) as an adaptation to island life. Paleontologists have long used this theory, known as the island rule, to explain both mammoth and miniature species found on islands. Flores hosted a community of dwarf stegodon, as well as giant rats.[68]

Yet, like the core theory of evolution, the island rule has never been proven. When researchers find an unexpectedly small or large fossil, they attribute it to the made-up rule. No one has found bones of normal-size stegodon or hominids on Flores. While *erectus* remains from 1.5 million years ago have turned up on Java, none have surfaced on Flores—and geological evidence shows the two islands have never shared a land bridge to link them. If these species shrank themselves *after* arriving on Flores, which the island rule requires, then where have the remains of their larger ancestors gone?

Perhaps, some say, *floresiensis* dwarfed before it arrived on Flores. Yet we run into the same problem here: how did they get to Flores?

Only one method existed for them. They had to cross the sea.

Troublemakers

Flores was hardly the first locale to stir up the anthropological pot. Since 1991, a site called Dmanisi, in the former Soviet state of Georgia, has yielded fossils of hominids aged between 1.7 and 1.8 million years old. For decades paleoanthropologists have nominated *Homo erectus* as the first world traveler, yet the bones from Dmanisi most resemble the older and more archaic hominid known as *Homo habilis*. The Dmanisi remains include the remnants of up to six individuals, varying from giants to petite creatures reminiscent of the newly discovered *Homo floresiensis*. The Dmanisi fossils have stirred up the pot so vigorously that the soup has slopped over the side. In fact, some paleoanthropologists lamented that maybe the bones should go back into the ground[32]—out of sight, out of mind, out of contention.

In typical paleoanthropological fashion, depictions of the Dmanisi hominids have softened over the years to placate the naysayers. The reconstructions sculpted shortly after the discovery depicted the Dmanisi hominids as apelike, hairy, primitive. More recent reconstructions have illustrated a softer, more humanlike version of the Dmanisi "people." Still far from human, these reconstructions nonetheless remove the shockingly apelike traits seen in early representations. They had to do it, of course. No ancestor of man, no member of the *Homo* line, could look so apelike.

The makeover has a hidden agenda. The Dmanisi hominids have been classified as *Homo erectus*, instead of *habilis*, in order to fit in with the notion of *erectus* as the first species to step out of Africa. Despite their affinities with *habilis*, their small brains, and the *habilis* tools found at the site, the first residents of Dmanisi must squeeze themselves into the *erectus* family. David Lordkipanidze, director of the Georgian State Museum, has suggested the fossils represent an intermediary between *habilis* and *erectus*.[28] The humanizing of the Dmanisi fossils, which include dwarfish specimens, reinforces the claim that the Flores hominids descended from *erectus*. Is it mere coincidence that the article featuring an easier-on-the-eyes Dmanisi hominid appears in the same issue of *National Geographic* as an article about the Flores hominids?

We need no conspiracy theory to imagine why the Dmanisi article directly follows the Flores article. The former buttresses the latter. Yet both rest on shaky foundations.

More finds in recent years have set paleoanthropologists to fuming. Mexican researcher Federico Solorzano has collected many ancient bones over the years. Recently, however, he has come to believe one bone in particular, a fragment of brow ridge, belongs to a *Homo erectus* who lived near Lake Chapala, west of Mexico City[54] *Homo erectus* died out at least 100,000 years ago, and modern humans aren't credited with reaching the Americas until 13,500 years ago. Researchers from the Center for the Study of the First Americans (CSFA) have supported Solorzano, though cautiously. With no corroborating evidence, in the form of stone tools or other remains, the discovery must linger in scientific purgatory. Yet, even if more bones are found that confirm Solorzano's analysis, will science accept the discovery?

Don't hold your breath.

Mexico has offered up yet more surprises since Solorzano's announcement. In summer 2005, geoarchaeologist Sylvia Gonzalez and her team presented a stunning new find to the world—human footprints at least 40,000 years old. This time, the find originated south of Mexico City. Scientists from Bournemouth University used 3-D laser scanning to reproduce the footprints, so as to avoid damaging the original *in situ* tracks. The resulting casts look stunningly human. Although Gonzalez and her team exploited multiple methods to date the footprints, their failure to present the findings in a scientific journal has garnered them criticism. Yet, according to *Archaeology* magazine, journals such as the venerable *Nature* have refused to publish the team's results.[25] One may find it difficult to announce one's findings in a peer-reviewed journal if those journals refuse to publish the results.

Other finds, many spearheaded by the CSFA, have corroborated the concept that modern humans set foot on American soil long before 13,500 years ago. The skeleton of Kennewick Man—a modern human found along a riverbank near Kennewick, Washington State, in 1996—is 9,000 years old. The date might seem uncontroversial, but the analysis of the remains has proven highly contentious, as archaeologists studying the bones noticed Caucasian features in the skull. Four separate Indian tribes swooped in to claim the remains as their ancestor, which would give them the right under US law to demand the skeleton's reburial without scientific study. After a protracted legal battle, in 2004 a federal appeals court upheld an Oregon judge's ruling that the remains are too old to have any demonstrable link to modern tribes. This year, scientists finally had their chance to begin a detailed study of the skeleton. So far, their analysis meshes with earlier assessments—Kennewick Man's skull shows Caucasian traits.[43] This finding, like the evidence from Mexico, suggests humans entered the Americas from other locations besides Asia.

The theory designated "Clovis-First" states that the first humans to venture into North America, who created stone points known as Clovis points, arrived between 15,000 and 10,000 years ago, depending upon whose model you accept. Some researchers fall back on the intermediate date of 13,500 years ago. Either way, these humans trekked out of Asia across the Bering land bridge, into Alaska, and down the ice-free corridor through Canada into the United States. Yet the Mexico and Washington finds refute the Clovis-First model.

If we had two lone examples of anomalous remains, perhaps we could write them off as flukes or misidentifications. Yet examples abound.

At a site known as Monte Verde, in Chile, radiocarbon dating has proven that evidence of human occupation extends back a thousand years before the earliest accepted date for the Clovis culture.[61] To claim the people of Monte Verde immigrated southward from the Arctic *before* they left Asia turns the study of human origins into a comedy, or maybe a time-travel fantasy. By no means other than the wishful thinking of archaeologists could Monte Verde's residents have journeyed down from the ice-free corridor.

Dennis Stanford of the Smithsonian Institution in Washington, DC, has spearheaded efforts to uncover the true history of human migrations into North America. If the first Americans hiked over from Asia, then we should find evidence of the Clovis culture, or a pre-Clovis toolkit, in Alaska. But Stanford searched Alaska and found no such evidence. When he and his colleagues examined tools from Asia, they found no evidence of a Clovis ancestor there either.[61] So from what mysterious place had Clovis emerged?

Expanding his search to Paleolithic cultures around the world, Stanford realized one pre-Clovis culture did create tools that share many traits in common with Clovis points. That culture, the Solutrean, left behind ample evidence of their stone-tool tradition in sites across France and the Iberian Peninsula. The Solutrean culture first appeared around 20,000 years ago, flourishing for 4,000 years. The Solutreans became the first people to shape their stone tools via pressure flaking, a technique that involves pressing on the stone with a piece of hardwood or antler in order to remove flakes. Pressure flaking shaves off long pieces of stone, flakes so long in fact they're comparable to the fluting technique used by the later Clovis people.

The Solutreans also used another technique straight out of Clovis—the overshot method, by which they would slice a single flake off the entire face of the stone. While Stanford has written countless pages about the likenesses between the cultures, one final parallel deserves mentioning here. Both Solutreans and Clovis people left behind similar carved tablets.[61]

And what about Mungo Man? Fossils found at the Lake Mungo site in southern Australia have made the jobs of paleoanthropologists more difficult, leading to conflicting, ever-changing theories about how and when humans entered the southern continent. Originally dated to 30,000 years ago, newer radiocarbon results have shown the human remains from Mungo date to 62,000 years ago.[62] The bones from Skhūl, dated to 80,000-100,000 years ago, supposedly represent modern humans, though they appear more Neanderthal-like. Even if we accept the Skhūl bones as modern human, then we have a span of no more than 38,000 years in which humans traipsed from Israel all the way down to Australia, half a world distant.

A skull found at Kow Swamp, Australia.

But Australia had hardly given up all her secrets at Mungo. In Kow Swamp, hundreds of miles south of Mungo, excavators dug up the remains of several "humans" whose occupation of the site dated back 10,000 years. What's so strange about that? The Mungo people, who lived 62,000 years ago, appeared quite human. At Kow Swamp, however, the "people" had hefty jaws, hefty teeth, hefty faces, sturdy brow ridges, and flat foreheads.[63] Hmmm...sounds familiar, doesn't it? A comparison to Neanderthals or *Homo erectus* sounds more than plausible. Unlike any hominid species, however, the Kow Swamp beings had severely elongated craniums.

Paleoanthropologists have clamored to concoct an explanation for the reverse evolution witnessed by the Australian fossils. Some say two populations of humans existed, somehow remained separate for 50,000 years, and then merged to become the ancestors of the modern-day aborigines. However, no evidence of earlier archaic hominids in Australia has been found.

Another theory explains that the people of Mungo were all women, the Kow Swamp people all men, and the elongated craniums with flat foreheads seen on the Kow men stem from skull binding[63]—the old standby excuse for any unusually shaped skulls, such as the elongated skulls found in South America. The theory goes that, as infants, these people endured one of two scenarios: they had their heads bound onto cradle boards, accidentally causing the elongation and flattening; or, they had their heads sandwiched between two boards in order to flatten and elongate the head on purpose as part of some ritual. Never mind that no one has ever seen a living person whose head has become seriously deformed by skull binding.

On my *Strange Origins* blog, I posted an entry about these deformed skulls and the mystery of who or what they represent. A reader posted a comment in which he claimed to have photos of living people with heads elongated by skull binding—as usual, though, the claim proved baseless, as he never provided the photos, even after I contacted him to personally request he post them on the blog. Anyone can say with authority "I have the proof." Offering up that proof for public inspection is another matter altogether.

The sole evidence that skull binding can deform heads comes from very old movie footage of people in Africa who have slightly elongated heads. Cradle boards can flatten the back of the skull, but not to a great degree. Besides, skull binding has no bearing whatsoever on the brow ridges, stocky jaws, or massive teeth of the Kow Swamp "people."

What does this leave us with? The silly scenario that the women, who apparently lived hundreds of miles away from the men, possessed human features and gracile bodies; the men, meanwhile, retained *erectus*-like features and robust bodies and, for some reason, decided to sequester themselves from their womenfolk. Maybe paleoanthropologists simply haven't unearthed the remains of the Kow Swamp women, or the Mungo men. Perhaps the men habitually crinkled their eyebrows, causing the brow ridges to form.

And maybe the ocean is salty because whales cry a lot.

The finds discussed above have elicited controversy, legal battles, and angst. Yet even older discoveries had cast a murky light on the landscape of human origins well before the sun struck the fossils of Dmanisi, Kennewick, or Lake Mungo.

No Humans Allowed

It's hard being a paleoanthropologist. You gather bones in the scorching African sun so you can use them to construct a pretty little timeline for human evolution. Then along comes a rebel who drives his motorcycle right over your timeline, smashing it to bits.

Fortunately, you can just ignore the rebel. You have the Darwin Police on your side.

Who are these rebels? Bones and tools, for the most part—artifacts sentenced to obscurity by the scientific establishment. Throughout the history of paleoanthropology, finds have emerged which, if accepted as genuine, would demolish the whole concept of evolution or, at the least, force us to question our ideas about how humans emerged and spread across the globe.

Humans evolved around 200,000 years ago. Science has posted a "no humans allowed past this point" sign in the history books. Yet the rebels defied the rules.

The Defiant Dead

In 1913 the German Hans Reck led a team to excavate at Olduvai Gorge in Tanzania, where fossils of extinct animals had been found two years earlier. The gorge contained multiple layers of sediment, with newer layers set down atop the older ones. Lodged in the wall of the gorge, in a layer approximately one million years old[20], Reck and his team discovered the skeleton of a modern human. Despite ridicule and disbelief from his colleagues, Reck stuck to his dating of the skeleton and contested the notion that the skeleton had become lodged in the rock via erosion and later, intrusive deposition. The skull was warped, a fact that seemed to indicate a long burial during which the deposition of new stratigraphic layers had exerted enormous pressure on the skull.[20]

Other paleoanthropologists complained that, if the skeleton was fully modern, Reck should have found stone tools in association with it. Originally, he found no tools. But when Louis Leakey instigated new excavations at Olduvai in the 1930s, he noticed stone tools everywhere at the site. Soon he realized where the problem had lain. Reck had failed to note the tools because he'd become accustomed to the flint tools found in Europe. At Olduvai Gorge, the tools were manufactured out of coarser rock from which only cruder tools could be fashioned.[64] Although initially skeptical, Leakey found Reck's reasoning persuasive enough to change his mind. He accepted the skeleton as an *in situ* burial of a modern human nearly a million years old.

Later, in the face of stern opposition, both Leakey and Reck would rescind their claims. Since both had studied the Olduvai formation in exquisite detail, leading them to rule out recent burial as the culprit, the explanation for their change of heart seems more likely to derive from professional pressure than factual error.

Consider this: Leakey also believed other remains his team unearthed at Olduvai, as well as Kanam and Kenjera[20], belonged to humans too, though he never managed to convince his colleagues. During his college years, Leakey had visited the Calico site in California, where he recognized stone tools 200,000 years old that must have originated with humans.[20] Yet humans, the dogma says, reached the Americas no earlier than 15,000 years ago. Leakey's colleagues and successors considered his forays into forbidden science an embarrassment. Yet they also consider Leakey one of the most brilliant stars in the paleoanthropology universe. Should we believe that an erudite man such as Louis Leakey would, throughout his career, persistently indulge in fuzzy thinking? Once, perhaps. But again and again?

If we had no more than Leakey's and Reck's discoveries to judge, ignoring them would prove easier. But the defiant dead have bequeathed to us more of their legacy.

In the 1970s, in Canada's Yukon Territory, archaeologists excavated a site known as the Old Crow River. There, in strata 80,000 years old, they dug up antlers and animal bones that exhibited tool marks on their surfaces, marks which had been made before the specimens fossilized.[20] Nothing aside from the human hand could incise tool marks on bones and antlers, and humans should not have inhabited the Yukon for another 65,000 years. Numerous sites around the globe have yielded similar tools fashioned from bone and rock in sediments so old they should bear no such tools—like the Red Crag formation in England. These strata date to 2-55 million years ago, and bone tools found within the formation showed signs of reworking by removing flakes across the grain of the bone.[20] Even a date of two million years ago flouts conventional wisdom, as no humans—no hominids period—should've lived in England during that era.

The Red Crag finds, along with plentiful examples of similar finds, happened in the 19th and early 20th Centuries, a time when questioning the newborn creed of evolution didn't brand a scientist a heretic or endanger his professional status. Sir Alfred Russell Wallace, who developed a theory of evolution by natural selection at the same time as Darwin, later examined the evidence for a human presence in the Americas during the Tertiary epoch, 5-65 million years ago. He found the evidence compelling enough to write an article on the subject, in which he decried the scorn many of his colleagues heaped upon the notion of man's antiquity. In particular, he mentioned the "Calaveras skull," a modern human skull found in sediments dating to the Tertiary.[20]

Even more ancient evidence turned up in northern Texas in 1969, when a group of Creationists undertook an excavation at Glen Rose, southwest of Fort Worth. The Films for Christ Association documented the excavations on film, as layers of limestone were peeled away to reveal the underlying rock. In the newly exposed rock, the team uncovered humanlike footprints. The limestone of Glen Rose dates back 100 million years to the Cretaceous Period. In the same limestone deposits, paleontologists find the fossilized tracks of dinosaurs. Some of the human tracks appeared alongside dinosaur tracks, and some even crossed over the dinosaur imprints.

In the 1970s, the tracks caught the attention of John Green, a Canadian Bigfoot researcher. In his seminal 1978 work *Sasquatch: The Apes Among Us*, Green documents his trip to Glen Rose, where he saw the tracks firsthand. He thought the tracks might represent Cretaceous hairy hominids, so he carried out a detailed investigation of the tracks. Though he later decided they were not the prints of Bigfoot, he accepted the authenticity of the finds. The tracks, he said, evidenced ridges where the foot had pushed the mud up as it sank into the muck. No paleontologists he spoke to knew of a dinosaur with narrow, 18- to 20-inch-long feet. Green also dismissed a hoax, since the film of the original excavations showed a crowd of people milling around the area.[33] How would the hoaxer have time to carve out the footprints? Besides, some of the prints had been found inside Dinosaur Valley State Park, in the middle of the Paluxy River, the park's main attraction.

While researching *The Hunt for Bigfoot*, I visited Dinosaur Valley State Park to look for the "man tracks" myself. I had learned of them via a website and wanted to see them for myself. Most of the best prints have eroded away due to river flooding, a fact the original excavators mentioned to John Green[33], but I had hopes of seeing some along a shelf about three feet above the riverbed. A drought had reduced the Paluxy to a trickle. I saw no obvious tracks in the riverbed itself, though I did stumble upon a vaguely foot-shaped depression that looked quite different than the depressions caused by water erosion, which I could see all around me. I never did find the shelf—but then the map I'd acquired online offered little detail. Disappointed, I stopped in at the Creation Evidence Museum just outside the park. I'd also read the museum housed several examples of fossilized footprints.

The Creation Evidence Museum occupies a doublewide trailer along the highway. Inside, displays market books about Creationism and related

A footprint-shaped depression in Cretaceous-Era limestone, cut out of the bed of the Paluxy River and relocated to the Creation Evidence Museum (Glen Rose, Texas).

While I failed to locate any footprints *in situ* at Dinosaur Valley National Park, I saw several footprints in the Creation Evidence Museum, just outside the park. Someone had applied a wash to the limestone to make the footprints more visible; however, the original shape is still obvious and enough like a human footprint to raise questions—how did a human footprint become fossilized in 100-million-year-old rock?

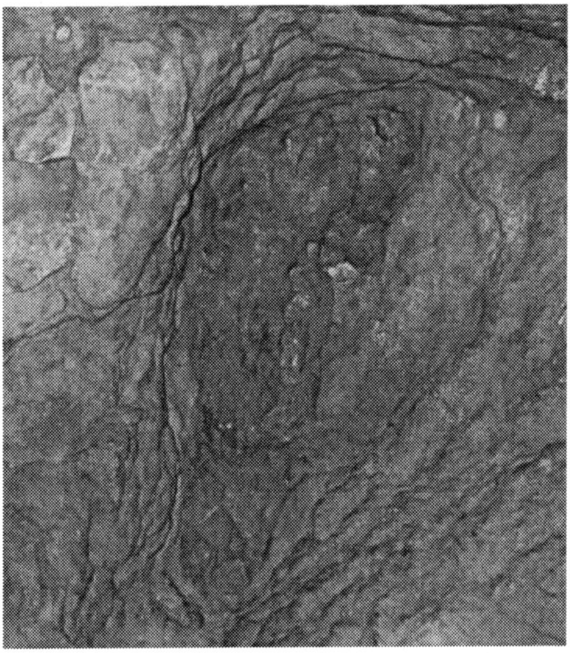

Another of the footprints now housed in the Creation Evidence Museum, this one apparently overlaid on a dinosaur track (Glen Rose, Texas).

themes, including the "man tracks." Toward the back of the museum, a hunk of rock sits propped up on a stand. The rectangle of stone holds a foot-shaped depression complete with suggestions of toes. While someone had applied a dark wash over the track to make it more visible, its shape would've been obvious without the visual aid. The depression possessed both the size and contours of a human foot.

Other tracks found at Glen Rose, like the ones John Green examined, are longer than a human foot. Ranging from 18 to 20 inches in length, they have toe marks and ridges where the sediment pushed up as the foot pressed into the mud. Green's main reason for thinking the tracks didn't belong to a hairy hominid centers around the width of the tracks, which he considered too narrow for a Bigfoot. A Northwest Bigfoot might have wide feet, but hairy hominids in other parts of the US—indeed around the world—have feet of differing dimensions. The hairy hominid footprints I found in Texas were narrower than the typical Sasquatch tracks from western Canada and the Pacific coast of the US.

The fact that Creationists brought the Paluxy tracks into the limelight has given scientists a good excuse to dismiss them without so much as glancing at a photo of the tracks. After all, any devout Christian must harbor an insidious willingness to do anything to further his or her cause (to prove humans did not evolve from apelike creatures). How would they explain John Green, a non-Creationist, who found the Creationists' arguments for the authenticity of the Paluxy tracks irrefutable? Scientists have become zealots of another religion, the Church of Darwin, to which they vow blind faith. Anyone who opposes them gets lumped into the Creationist category, a euphemism for "whacked-out nutcases."

I never did see the best footprints from the Paluxy trackways in person, though I saw photos of them on the Creation Evidence Museum website. The mystery intrigued me—in fact, it still does. My visit to Glen Rose persuaded me to start my novel, *The Hunt for Bigfoot,* with a scene at Dinosaur Valley State Park in which the heroine discovers human footprints fossilized in the limestone riverbed. I am not a Creationist, but I refuse to discount evidence simply because a person with deep religious beliefs uncovered it.

One need not become a Creationist in order to question the prime tenet of human origins. Creationists believe God created the universe out of nothing, and later fashioned humanity in His own image. The truth may lie somewhere between the two extremes. The theory of "intelligent design" is similar to Creationism in that it requires a supreme being, yet removes some of the religious overtones from the argument. Interventionism, a newer theory for the origin of life, accepts that beings from another world (or galaxy, or perhaps another universe/dimension) seeded life on Earth.

You might be surprised to know that most astronomers now believe life *must have originated elsewhere* and been seeded on Earth by some external force. They consider an asteroid the most likely source for life-seeding organisms. Yet the very fact that they recognize life could not have sprung up here without assistance is telling.

Molecular Mischief

Paleoanthropologists have one last card to play. They claim to have proven, beyond doubt, that modern humans evolved in Africa about 200,000 years ago—from a single pair of ancestors, nicknamed Adam and Eve. The claim sounds outlandish. Unless you read a great deal about the DNA evidence, unless you take the time and effort to compare the differing opinions from scientists, the proclamations of paleoanthropologists seem convincing. If you brush away the surface dust, however, the true nature of the DNA evidence becomes clear.

Finding Eve

Every human being has two types of DNA in his or her body. One type, nuclear DNA, resides in the nuclei of our cells and consists of equal parts of the DNA from our mothers and fathers. Mitochondrial DNA (mtDNA) resides outside the nucleus in energy-packed entities called mitochondrial organelles. Unlike nuclear DNA, mtDNA comes down to each of us as a gift from our mothers, since males contribute no mtDNA to their offspring. Thus, mtDNA preserves markers from all our female ancestors. While scientists have studied nuclear DNA, it has proven more problematic than mtDNA. Nuclear DNA, inherited from both parents, undergoes arbitrary reshuffling that renders it unreliable as a data source.

How far back can mtDNA take us in our lineage? Paleoanthropologists would say all the way back to the beginning. Yet the answer to the question is hardly as cut-and-dried as they'd have us think.

Our mitochondrial DNA first stepped into the spotlight back in 1987. Scientists from the University of California at Berkeley had taken samples from the placentas of 147 women of diverse ethnic makeup, then analyzed those samples to find the mtDNA present in them. Since mtDNA contains remnants of our female ancestors' DNA, they believed they could trace humanity—the female branch of it anyway—backward in time through genetic markers. In 1987 they announced they had found "mitochondrial Eve," the mother of every human, one woman who had spawned a species. The team claimed to have found her homeland too, in Africa. To accomplish that feat, they had utilized mutations.

Like a computer file copied onto a floppy disk hundreds of times, overwriting the old copy with each repetition, mtDNA becomes corrupted over time. Biologists call these corruptions "mutations." The Berkeley team exploited mutations to backtrack our mtDNA for 200,000 years, to an epoch when the dogma states modern humans had recently evolved. Therefore, the mtDNA must be tracking human ancestry back to the first modern humans. Since people of African descent have a higher rate of mutations, hinting at a higher genetic diversity, humans must've originated in Africa. Biologists assume that higher genetic diversity equals an older population. But the Berkeley team relied on a series of assumptions in order to assemble the jigsaw pieces into a recognizable picture.

First, they assumed mutations occur at a regular rate that can be both measured and predicted. But many factors can influence, and in fact distort the appearance of, the mutation rate. Four bases—A, C, G, and T—comprise our DNA, like an alphabet that spells out instructions for our cells. Remember the floppy disk analogy? The mutations occur when the sequence of these bases becomes corrupted. An A mutates into a T, or a C replaces a G.

In analyzing the mtDNA of those 147 women, the Berkeley scientists assumed mutations occurred at regular, measurable intervals. However, mutations can occur spontaneously, for no apparent reason. Our genes don't care what scientists think. They'll mutate when the mood strikes them. As for the measurability of mutations, that remains questionable too. If an A switches into C then, much later, decides to become an A again, scientists would *fail to see that any mutations had occurred.* The A would appear as if it had always stayed an A. Its changes of heart would be undetectable. What happens if an A pushes aside a G, only to get swapped itself for a C generations later? The bases have mutated twice, but geneticists will see but one mutation.[63] Returning to the floppy disk analogy, if you overwrite a file twice, you won't be able to open the original file. You won't even be able to see it. The file will appear to have changed once, as the "last modified" date attached to it will reflect only the most recent change. A similar situation happens with our mtDNA.

The second assumption involves computers in more than just a metaphorical sense. The Berkeley team used a computer program to process the data they'd collected. The program dutifully swallowed the data and spat out a family tree for humanity. Yet, as other geneticists pointed out, the same data could impart different trees depending upon what software is used to process it and how many times that software analysis is run. Thousands of possible family trees lurk inside the data collected by the Berkeley team.[63] Which version you choose depends upon what outcome you desire. For instance, another scientist noted how the Berkeley team had, in two separated papers published in 1991, listed two different populations as the root ancestors of all humanity. In the first paper, the team listed Kung bushmen as the root; in the second, pygmies occupied that slot.[63] Both could not be true.

The third problem with the Berkeley study has its origins in the way populations grow, spread, and contract over time. The Berkeley study involved 147 women alive today. The sampling is static. Human beings, however, are not. We travel, we move our homesteads, we intermingle with other races. Your mtDNA won't show if your father came from Germany and your mother from Canada. Only your mother's mtDNA survives in yours. All your male ancestors get erased from the equation. Then we have the issue of population changes, for population bottlenecks can squeeze the diversity right out of human genes. As recently as the Middle Ages, the bubonic plague killed one third of Europe's population.

Some evidence suggests a bottleneck that occurred between 27,000 and 53,000 years ago could have reduced the population in northern Europe to j15,000 individuals or fewer. The bottleneck must have caused a matching contraction in the genes of northern Europeans. Hence, no one knows whether current genetic differences stem from a recent African origin of humanity or the bottlenecks which afflicted European populations.

Yet another quandary arises from the test subjects selected for the study. The Berkeley team had obtained ⅔ of their samples from women in American hospitals, and just two of the "African" subjects had actually been born on that continent. And what about the basis for the mtDNA clocks? Scientists needed some piece of data with which to calibrate their genetic clock. They opted for the date when "humans" split off from the chimpanzees.[63] But *no one knows* when (or, indeed, if) this split happened. Current estimates range from four to eight million years ago, though a few decades ago the estimate stood at 20 million. By presuming humans share a common ancestor with chimps, and that current beliefs about when the split happened are accurate, the scientists adjusted their mtDNA clock to fit their preconceptions.

Even scientists who support the Eve theory admit, if only in private conversations with other paleoanthropologists, that the mtDNA evidence shows Africa just barely nudges ahead of Asia in the calculations of the most likely birthplace for humanity.[16] They could as easily have found an "Asian Eve."

While other criticisms of the Berkeley study have emerged, the handful elucidated above emphasize the weaknesses of the "African Eve" theory. In response, scientists have sought other kinds of genetic evidence to shore up the framework of the "out of Africa" theory.

Nuclear (Non)Reaction

As I mentioned earlier, nuclear DNA has proven more difficult to analyze because of the disorderly manner in which it gets reshuffled, which has its roots in the fact that we inherit our nuclear DNA from both parents. But this little problem hasn't stopped scientists from drafting nuclear DNA into their evolution army.

A group at Yale University focused their nuclear DNA studies on one part of the strand—chromosome 12—and the sequences of base pairs found in two sections of that chromosome. They pronounced that, while Africans have much diversity in those sequences, non-Africans stick to one pattern. This, they asserted, proves once again that humans evolved in Africa and spread outward from there. They arrived at a date of 100,000 years ago for the moment when *Homo sapiens* departed Africa. [63]

Yet the problems with this study echo the criticisms of the "African Eve" study. The random shuffling of the nuclear DNA turns this area of study into a muddy swamp. Population bottlenecks could've reduced the diversity in non-African genes as well.

Another study, also at Yale, probed a segment of the Y chromosome, a section of nuclear DNA which we inherit solely from our fathers, much like our mothers hand down mtDNA to us. This time the scientists elected to study just eight men—a miniscule sample given that billions of people inhabit the Earth today. From the tiny sample they inferred that "nuclear Adam," father of us all, lived 270,000 years ago.[63] Hmm... sounds like he and "Eve" missed each other. Subsequent studies have elicited younger dates, pointing up the same old problem—the date you obtain depends upon what software you use and what timeline you wish to see.

Geneticists around the world continue to tease dates from our DNA, yet every attempt rests on the presumption that all humans derived from Africa. After all, as paleoanthropologists will gladly tell you, we know *Homo sapiens* evolved in Africa. We need only prove when that event transpired.

Too Many Eves

In the book *Other Origins,* anthropologist Russell Ciochon and his coauthors spotlight other evidence that contradicts the out-of-Africa theory. First, fossil evidence demonstrates that modern humans dwelled in Europe and Asia *before* the date now accepted as the time of our birth—200,000 years ago. Second, no specimens of *Homo sapiens* indisputably dated to 200,000 years ago have cropped up in Africa.

The central issue, however, remains the question of what happened after the "first humans" left Africa. The out-of-Africa theory requires that humans displaced the *Homo erectus* already living in Europe and Asia, without mixing with them.[16]

Berkeley's African Eve has acquired some company too. A study of human teeth, both from living and fossil specimens, has suggested modern humans arose in Asia. Just as the Eve study relied on statistical analysis of mtDNA, the dental inquiry rests on statistical analysis of how the specialized features on human teeth evolved and assumes the mutations that caused the specializations originate with ecological stresses brought on by our species populating new areas of the globe. The fewer the specializations exhibited, according to this study, the older the population. The Asian teeth showed the fewest specialized features.

Adding to the confusion, research in Asia has gathered profuse evidence that primates originated there, not in Africa.[16] So now African Eve must duke it out with her Asian twin to win the title "Mother of Us All."

Without a consensus on either the timeline or the fossil evidence, and with DNA data that must be distorted to fit the facts, any conjecture about human origins remains dubious. If the ancient hominids didn't mature into modern humans, evidence of their true descendants must exist somewhere.

The Essence of the Matter

Boiling the evidence down to the essence gives us several points to contemplate:

- ✧ The ancient hominids resembled humans in only the most superficial ways; they bore more resemblance to modern hairy hominids.
- ✧ Paleoanthropologists can neither accurately define what makes a species nor determine relationships among extinct species.
- ✧ Paleoanthropologists declare each new bone a new species of hominid, turning the hominid family tree into a mess of brambles.
- ✧ New species pop up suddenly, and vanish with equal abruptness.
- ✧ To support evolution, paleoanthropologists must distort the fossil evidence and ignore specimens that don't fit into the theory.
- ✧ Credible evidence of a human presence hundreds of thousands, even millions, of years ago exists—along with 100-million-year-old tracks that resemble the prints of modern hairy hominids.
- ✧ DNA studies, rife with flaws, have failed to fend off questions about evolution and human origins.

Where does the answer lie? Perhaps it has hidden right in front of us, masked by the forest.

4
Mistaken Identity

For about 700,000 years, beginning one million years ago, a giant ape lumbered around Asia. This extinct creature, known to science as *Gigantopithecus*, today acts as a scapegoat for a mystery that stymies those who examine it. Cryptozoologists say Bigfoot represents a descendant of *Gigantopithecus*. Nearly every organization dedicated to Bigfoot research, especially on the American west coast, subscribes to the "Giganto theory."

And why not? A giant, bipedal primate who stood 9-12 feet tall and sported a massive jaw that dwarfs that of a modern gorilla...sounds good. Too good. Anyone who has watched a television documentary about Bigfoot has likely seen a reconstruction of *Gigantopithecus*—a statue of a colossal ape, usually displayed with a man alongside it to highlight the enormity of the beast.

Such reconstructions give the impression that anthropologists have unearthed the complete remains of at least one member of the species *Gigantopithecus*, that we know exactly how the creature looked in life, and so we can say with certainty that it looked much like Bigfoot. Yet the truth belies the implication.

The facts simply don't support the hypothesis. Bigfoot researchers have latched onto the theory because it sounds reasonable, superficially, and provides a nice cushion for them to fall back on when confronted by the "lunatic fringe"—you know, those whacked-out potheads who insist on bringing up the "paranormal" aspects of the phenomenon. When facing evidence of a UFO connection, the researcher can simply say, "But we've proven Bigfoot's just an ape. Look at *Gigantopithecus*."

Well, let's look at old Giganto.

Dragon Teeth

The first hints that a giant ape had once roamed Asia cropped up just before World War II. The German paleoanthropologist Ralph von Koenigswald had begun searching Chinese pharmacies for fossils, which the locals called "dragon teeth" or "dragon bones" and used in homeopathic remedies. The Chinese pharmacies had earlier aided him in his quest to find remains of *Homo erectus*.

In 1935, von Koenigswald stumbled across a tooth he surmised belonged to a colossal ape. Further hunting netted him three more teeth. On the basis of these four teeth, von Koenigswald declared a new species of extinct ape that he christened *Gigantopithecus*, or "gigantic ape."

Another scientist, Franz Weidenreich, studied casts of the fossil teeth and reached a different conclusion. Noting the curious bulk of some *Homo erectus* fossils found on the Indonesian island of Java, he theorized that the teeth von Koenigswald had discovered belonged to a giant version of

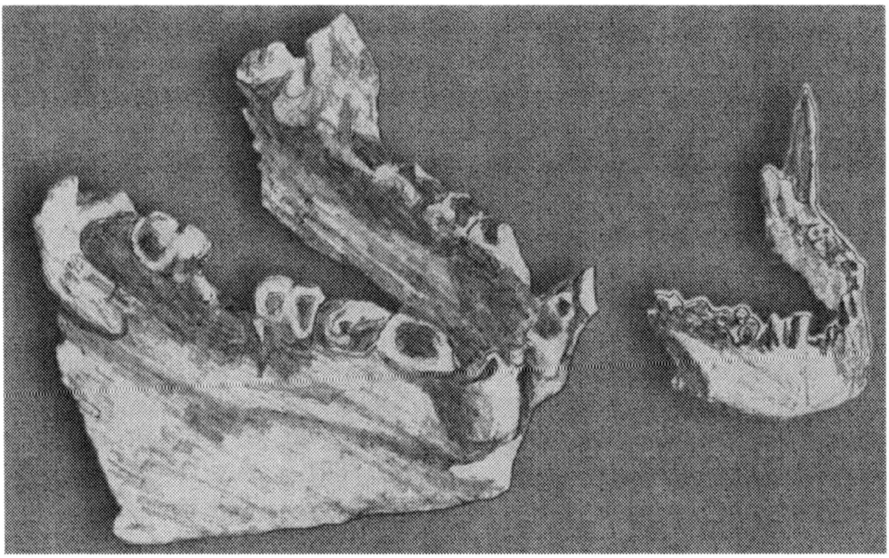

A *Gigantopithecus* jaw (left) in comparison with a human jaw (right); jaws and teeth comprise the sum total of the evidence concerning *Gigantopithecus*, the famed giant ape that died out 300,000 years ago.

Homo erectus, from whom modern people had evolved through a process of gradual shrinking.[17] As paleoanthropological theories go, Weidenreich's was no more fanciful than most. In fact, for years afterward, his theory supplanted von Koenigswald's vision of a giant ape.

In the 1950s, Chinese paleoanthropologists initiated a new search for the giant ape. Tracking down the source of the fossils in the pharmacies, the Chinese researchers located a cave high in a natural rock tower—a site which yielded teeth identical to the ones von Koenigswald had found 20 years earlier. At a site further to the north, another member of the team learned of a farmer who'd found a giant jawbone. Excavations at the same cave where the farmer had found his treasure yielded another jawbone. The new jaw retained most of its teeth, with just three missing. Now the paleoanthropologists could compare the teeth already found to the ones still affixed to the jawbone, which proved to them that the original specimens did indeed belong to a giant ape. Fragments of a fourth jawbone turned up in India in 1968, proffered by a farmer who had held onto the specimen for 24 years after happening upon it in a field.[17]

To date, in addition to the jawbones, paleoanthropologists have also collected about 1,000 teeth. No more bones of *Gigantopithecus* have surfaced. Though reconstructions of the beast imply we have a complete skeleton from which to work, in reality artists build their conceptions of the giant ape from four jawbones, some teeth, and a hefty measure of imagination. Even Russell Ciochon, one of the top experts on *Gigantopithecus*, admits in his book *Other Origins* that any reconstruction of a fossil animal involves at least as much art as science and produces only one of many plausible visions of the creature.[16]

Ciochon and his colleagues commissioned one of the several reconstructions in existence. Rather than contracting a scientist or forensic sculptor, they joined forces with a Hollywood special effects guru, Bill Munns, who also constructed models of living primates. The model Munns created for Gigantopithecus stands 10 feet tall. Critics of the model have observed that the reconstruction uses the largest possible dimensions for the creature[50], rather than a median size. Munns' model also resembles an orangutan, while other models take their cues from the gorilla.

All reconstructions of *Gigantopithecus* rely on conjecture, since the entire catalog of *Gigantopithecus* fossils consists of four jawbones and a heap of teeth. Scientists must estimate the skull size by comparing the jawbones to those of living apes, as well as other fossil apes. Then they must guess at the head to body ratio in a similar manner. Estimates generally range from 9 to 12 feet for height, 600 to 1200 pounds for weight. Yes, this still gives us a giant ape. It must be Bigfoot then, right?

Not so fast. The majority of experts who have studied the jawbones, including Russell Ciochon, believe *Gigantopithecus* preferred "knuckle-walking"—a quadrupedal gait—to bipedal locomotion. Only one primate alive today walks upright exclusively: humans. Other primates may walk bipedally on occasion, but spend the vast majority of their time either swinging from branch to branch or knuckle-walking. The gibbons and siamangs of Asia, for instance, adopt bipedal locomotion on occasion but swing amongst the trees 90% of the time.[49] Most Bigfoot researchers, a troop that includes a handful of biologists and anthropologists (like the late Grover Krantz), seek reasons to think *Gigantopithecus* walked upright. They mention how the jaws of *Gigantopithecus* spread out toward the back in a manner consistent with the shape seen in bipeds' jawbones.[39]

Yet comparing the *Gigantopithecus* jaw to those of living apes and humans points up the similarities to ape jaws, and the disparity with the mandibles of humans and the ancient hominids. Human teeth sit in a parabolic curve, while ape teeth align in two parallel rows. Ape jawbones have a V shape, while human jawbones have a more rounded appearance. The jaws of *Gigantopithecus* have a V shape, in line with living apes. Relative to the other teeth, the canines are much larger, another trait seen in modern apes. The Giganto jaws are also much deeper than human jaws.

The closest comparison found in the hominid fossil record appears in the jaw of *Australopithecus anamensis,* a species which dates to around four million years ago, and whose jaw bears a striking similarity to those of chimpanzees[42]. But do the fossils labeled *Australopithecus anamensis* truly belong to a bipedal creature? Paleoanthropologists claim the leg bones found nearby suggest *anamensis* did walk upright, yet no one can say for certain the jaw and the leg bones came from the same individual or even the same species. Collecting fossils by finding fragments of skeletons scattered on the ground leaves immense room for error.

Besides, if we accept *anamensis* as a bipedal hominid, then we have a problem. Its supposed successor, *Australopithecus afarensis,* the species from which Lucy hailed, has a jawbone and teeth that resemble those of humans far more than those of other primates. From the form of the teeth to the shape of the jaw and alignment of the tooth rows, *afarensis* displays its affinity to the only other creature whose bipedal nature remains beyond question—human beings. Yet, as discussed in the previous chapter, the ancient hominids diverge from us in too many other ways for us to have evolved from them, and further evidence suggests humans have existed for hundreds of thousands or perhaps millions of years.

Australopithecus anamensis seems a better candidate for ancient chimp than for ancient hominid. The resemblance between *Gigantopithecus* and *Australopithecus anamensis* jawbones denotes their affiliation with primates rather than with humans or the ancient hominids. Primates are not bipedal. Yet, to stand as the ancestor of modern hairy hominids, Giganto must have walked upright. To name *Gigantopithecus* as the ancestor of hairy hominids requires we accept that, while no other primate before or since has adopted a bipedal gait, for some reason Giganto did. Apes enjoyed their knuckle-walking status for millions of years then—boing!—*Gigantopithecus* pops out of the evolutionary box as a biped. In spite of its bipedalism, Giganto retains its apelike jawbone and teeth. Since we have no other bones of this creature from which to deduce its appearance, we can do no more than imagine how it looked. Imagination provides fertile soil for growing theories. Unfortunately, none of the ideas that sprout up have enough strength to mature.

Not one of the reconstructions of *Gigantopithecus* resemble Bigfoot anyway, save for the giant stature. Each reconstruction takes its cues from the ape world, which seems prudent considering the similarities between the jawbones of Giganto and other primates. In contrast, witnesses who've seen hairy hominids mention the humanness of the creatures' faces. Their testimony also provides other clues that simply don't mesh with the Giganto theory.

Anatomy of a Giant

A major anatomical feature that distinguishes humans from apes is our opposable thumb—opposable meaning we can move our thumbs under our palms to grasp things. Few witnesses have seen hairy hominids at close enough range and for a long enough time to catalog the physiology of the creature. Of those who have, fewer still notice the thumbs. One witness who saw two creatures handling rocks said that he couldn't see the thumbs, because either they didn't have any or they didn't use them.[33] We should refrain from assuming that if hairy hominids had opposable thumbs they would use them the same way we use ours. They don't live like us, so why should they behave exactly as we do? Perhaps, to them, another method of handling rocks seems more sensible—or, perhaps, the witness simply had a bad angle for viewing their thumbs.

In *Bigfoot Sasquatch Evidence,* Grover Krantz discusses several casts of handprints reportedly made by hairy hominids. Based on the casts, he believed the creatures had non-opposed thumbs. However, the casts offer nothing more than vague clues to the nature of Bigfoot thumbs. Krantz describes how the position of the thumb in the casts indicates a non-opposed digit, as an opposable thumb would lie further back from the other fingers.[39] While his observations about footprints are excellent, here he seems to stretch the evidence in order to support his belief hairy hominids descended from *Gigantopithecus.*

Two of the handprint casts were obtained by a man widely believed to have faked Bigfoot evidence. Of these, one shows the fingers either pressed tip-down into the soil as if the fingers had been bent, or with the knuckles folded. Either way, the cast offers little evidence on which to establish the position of the thumb. The other cast is so indistinct that even making out the fingers takes much scrutiny. Two other handprint casts, obtained 14 years later, show slightly more detail. Yet these casts also appear distorted, with the fingertips flattened and the position of the thumb difficult to ascertain. The cast of the right hand appears complete, and presents the most detail, though it also seems distorted. A fifth cast includes only the knuckles and an apparent thumb. To judge any creature's anatomy based on indistinct casts begs criticism. At least with the footprints we have numerous examples from locations all over the world, a catalog which includes clear impressions, not just vague shapes.

In truth, inferring anatomy from casts puts the researcher in a very precarious position. The process of casting can deform the track, especially in soft soil, altering the evidence inherent in the track itself. Researcher Archie Buckley, an expert in human foot anatomy who studied Bigfoot tracks in California, wrote about this issue way back in 1984. From his experiences, he'd concluded that the same hairy hominid could, in the space of a single trackway, impress his feet differently with each track due to changes in the terrain and the speed with which he moved, as well as his balance. So, the footprints in one trackway could vary in length and width, depth of impression, and apparent anatomy. Thus, casts that appear to have no arches may indeed have been made by a foot with a humanlike arch.

Buckley felt that variations in the resistance of the soil could also affect how the researcher perceived the anatomy of the foot that impressed the track. Therefore, tracks may exhibit no arch because the soil's hardness didn't allow the foot to impress deeper as the arch came down onto the soil. The creature might also trip, slip, or otherwise distort the prints when walking down or up a hill, traversing a slippery patch, or fleeing at high speed. Buckley states that his studies show *no anatomical differences between a human foot and a Sasquatch foot except in how they use their toes.* They do, he believes, use their toes more actively than humans—the main reason being that we wear shoes, which limit our toe movements, and they do not. Buckley discusses finding, in a single trackway created by one individual, both footprints that appear flatfooted *and* hourglass-shaped tracks.[10]

Buckley presented no specific evidence that hairy hominids have arches in their feet. Nonetheless, his experiences do point up the problems with relying on footprints and/or casts of footprints in determining a creature's anatomy. The Bigfoot captured in the Patterson-Gimlin footage seems to have flat feet—which does not mean all hairy hominids have flat feet.

I have seen footprints that could support Buckley's assessment. One print I found in Texas included an indentation along one side. At first glance, it would appear as if a portion of the print had been brushed away, or perhaps another animal had stepped on the track, obliterating part of it. However, closer scrutiny revealed no animal tracks overlapping the hairy hominid track. What other mechanism could've erased part of the track? None that I can find. Another explanation seems pertinent and, indeed, far more probable. The indentation marks the arch of the foot. Another track, this time from Michigan, appeared to belong to a juvenile hairy hominid. That track looked human, except that no humans live in the gloomy swamp behind my home.

Would a hoaxer think of creating a fake trackway that comprises both flatfooted and hourglass tracks? Probably not. To do so would likely convince the average Bigfoot researcher that the tracks were indeed fake. Some researchers subscribe to the idea that smaller Bigfoot tracks which fall into the size range of human feet *must be discounted.*[33] Such beliefs derive from a twofold idea: that any Bigfoot track which compares in size to human tracks would prove tough to distinguish from human tracks; and, that excluding human-size tracks makes Bigfoot researchers appear more reasonable. Yet, no matter how deeply we bow to the mainstream scientists, they will never accept hairy hominid research as legitimate or valuable.

Why should we care whether they accept our research or not? Their denials and debunking attempts do little to hinder the research—unless we let them discourage us. If a human-size track turns up in an area where no human resides or where no human would walk barefoot, and if the track bears hallmarks of a nonhuman origin, study it. If you find *any* size footprint made by a bipedal creature with a humanlike-yet-not-human foot, study it. Period.

Getting back to the handprints, similar issues arise with casts of handprints as with casts of footprints. First, we must wonder how the creatures left their handprints in the ground. A few witnesses have seen hairy hominids drop down onto all fours; such cases encompass a tiny minority of the sightings. The creatures may well brace themselves with their hands when negotiating a steep slope, or throw their hands out to retard a fall when they trip.

The same factors Buckley mentioned in association with footprints could play a role in the appearance of handprints too. The hardness of the soil, the grade of a hill, the slipperiness of the mud or sand—these aspects and others could disfigure a handprint or instill in it false impressions that lead the researcher astray. Hence, with imprints of any kind, we must take care to examine the entire length of the trackway and scrutinize each imprint, recording deviations amongst the individual tracks. We must also examine the surroundings to determine whether sloping terrain, hard soil, or mud influenced the appearance of the footprint or handprint.

The incomplete nature of the handprint evidence provides no closure on this issue. The evidence from sightings proves equally imperfect, since few witnesses notice the creatures' hands. Those who do either can't see the thumbs or can't tell if the thumbs are opposable or not.

Sightings do provide clues, though, concerning the hairy hominids' ancestry. Along with other types of evidence, the data intimates an entirely different genesis for hairy hominids, one closer to our own origins than we like to imagine.

A Basic Description

Sightings afford a unique window into the world of hairy hominids. While some witnesses may be mistaken and others may lie outright, many more tell credible stories of encounters with hairy bipeds. The descriptions given by witnesses maintain a stunning consistency throughout time and across geographic boundaries. From the sightings data, we can glean a basic description of the suspect.

They have arms longer than humans', with hands instead of paws. A few witnesses have reported "claws," but others have said the fingernails were sharpened into points, which would lend them the look of claws. Hairy hominids tend to have broad shoulders. Their heads rest on their shoulders so that they seem to have no neck. Hair swathes their entire bodies, save for the palms, the soles of the feet, and most of the face. Hair on the shoulders and torso tends to grow longer than the hair on the remainder of the body. The nose is flat and wide, as are the lips. They display brow ridges.

Texas Bigfoot are generally slender, while Pacific Bigfoot have stocky frames. Pacific Bigfoot also tend to have heads that taper into a cut-off cone. Other hairy hominids have rounded heads. Variations in physical descriptions from witnesses around the world could easily signify diversity within a species, rather than a symptom of unique species. Many hairy hominids have five toes, but three- and four-toed variants occur in some regions. In colder regions, frostbite could cause a number of the creatures to lose toes. Genetic anomalies likely account for the rest. After all, humans have differing sizes and shapes of feet. Some have flat feet, others very high arches. Some have missing toes, others extra toes. Certain people sport long, thin toes; others have short, squat toes. The list could go on for pages. Given the diversity of physical traits among humans, we can accept similar diversity among hairy hominids.

Witnesses frequently describe hairy hominids as humanlike. Phrases such as "I knew it was human," "it had a human face," and "it looked like a hairy man" appear in countless sighting reports. The long arms and brow ridge harken back to the ancient hominids. *Gigantopithecus* fits nowhere into this depiction.

The deviations in appearance also mirror the ancient hominids, whose fossils exhibit great diversity in physical appearance. Paleoanthropologists divide the hominids into separate species based partly on the differences in appearance (morphology), but the main impetus for declaring a new species arises from each paleoanthropologist's dream of discovering a brand new species. In reality, the anthropological world has reached nothing close to unanimous agreement on the issue of what defines a species, especially in the fossil record. All the ancient hominids may well belong to one species. The similarities between them imply they do.

The same reasoning, or lack thereof, has inspired cryptozoologists to partition hairy hominids into separate species. The American Bigfoot must represent a different species than the human-size Almas of the

Caucasus region in Asia, and the more apelike orang pendek of Sumatra must come from a different species than the Himalayan Yeti. Why? They look slightly different—Bigfoot are taller and stockier.

Yet fingerprint expert Jimmy Chilcutt, speaking on the radio show Coast to Coast AM with George Noory, stated that his analysis of hairy hominid footprints led him to conclude they all fit under the umbrella of a single species. Their footprints show too many similarities for them to represent distinct species.[5]

Since we have no *Gigantopithecus* footprints for comparison, we must rely on other evidence to reach a conclusion. We have more than appearances, though, on which to found such a conclusion. We must also consider behavior.

Act Like An Ape!

Nobody told the hairy hominids they were apes. Since the dawn of their species, they have behaved in decidedly un-apelike ways. Researchers who want to plant them in the primate garden never asked the hairy hominids how they feel about it, maybe because they know they wouldn't like the answer. We can yell "act like an ape, you insolent dolt!" until we pass out from oxygen deprivation, but the hairy hominids will refuse to obey.

In what ways do they act un-apelike?

Apes don't swim, for one. Hairy hominids do. In the 1830s, reports emerged from the area around Fish Lake, Indiana, of a four-foot-tall "wild child" loitering in the vicinity—and swimming in the lake.[7] In another incident that took place in September 1967, a fisherman casting his net on the delta of the Nooksack River in Washington State felt something tug on his net. A moment later something began dragging his net upstream. When he shined his flashlight at the thief, he saw a hairy hominid in the river hauling in the net. The witness hollered, which drew two nearby fisherman to him. All three watched as the creature waded toward a nearby island, where it vanished into the trees. Witnesses observed the Lake Worth Monster of Texas swimming to an island, while the Sister Lakes Monster of Michigan left tracks that disappeared into a pond.[33] In 1970, at Priest Lake in Idaho, a hairy hominid swam across the lake in full view of two fisherman. The creature kept its arms underwater, so the men couldn't tell what method of swimming the hairy hominid used.[7]

Maybe Bigfoot can dogpaddle...hey, you never know. If you want to limit the wake you stir up, so as to avoid detection as much as possible, dogpaddling will serve you well.

Maybe we should rename it Bigfoot-paddling.

The un-apelike behavior doesn't end with swimming. Hairy hominids also hunt. Sightings and research indicate hairy hominids feast on deer more often than any other animal, possibly to obtain the nutrient-rich liver[44], although they also seem to enjoy rabbits, chickens, turkeys, and perhaps even cows. In *The Bigfoot Casebook*, Janet and Colin Bord list 1,000 sightings that occurred in the years 1818-1980. The listing includes many instances of hairy hominids chasing and/or killing animals, as well as transporting or consuming the carcasses.

In 1885, in the Cascade Mountains of Oregon, several hunters saw a hairy hominid eating raw deer flesh. When the creature spotted them, it fled. More recently—in the autumn of 1957, at Wanoga Butte in Oregon—hunter Gary Joanis had just shot a deer. Before he and his companion, Jim Newall, could reach the carcass, a 9-foot creature snatched it off the ground and absconded with the prize. Joanis shot at the Bigfoot, to no avail. In November 1963, three men driving near Johnson Siding, South Dakota, were shocked to see a 10-foot-tall white Bigfoot caught in their headlights. Nabbing a dead deer from the ground, the creature fled into the underbrush nearby.[7]

A man named Herb Brown recounted how, while in California in the mid-1960s, he observed as a Bigfoot stalked four deer. A century earlier, in northern Nevada, an armed party pursued a hairy hominid who wielded a club in one hand and a dead rabbit in the other. They lost their quarry, but came home with an incredible story.[7] Lest the reader think this case represents an anomaly, here's another case involving a hairy hominid brandishing a club. On January 14, 1902, a group of young people skating on a pond near Chesterfield, Idaho, drew the ire of a hairy hominid. Hollering and waving a club, the creature charged at the youngsters, who escaped into the safety of nearby wagons.[7]

In 1994, a family driving through the woodlands of central Wisconsin had a near-miss with a hairy hominid crossing the road. The creature carried a dead deer over its shoulder. In 1992, near Wills Creek in Ohio, a man hunting in an isolated area happened upon a strange site. Under an overhang, in a small grotto, he discovered a number of severed deer legs arranged in rows. The hunter decided to take a break there—until the owner of the deer legs returned. The hairy hominid simply stared at the intruder until the man finally stole past the creature and out of the cave.[44]

In March 1851 in Greene County, Arkansas, a witness first watched a herd of cattle stampede past him, then spotted what had spooked them—a hairy hominid in pursuit of the bovines. The creature paused to study the witness before taking off again.[33] In those days, newspaper reports often dismissed the creatures as men who had become stranded in the woods and "gone wild." Somehow the act of abandoning civilization had caused these "wildmen" to grow a coat of hair all over their bodies! Hmm...guess I'm in trouble. Living here in the remote woods of Michigan, removed from "civilization," I may soon acquire a covering of hair.

After the discovery of mountain gorillas in Africa, the explanations shifted from men gone wild to wild apes. Any hair-covered being who looks similar to humans must be an ape, the reasoning goes. But the evidence suggests the contrary. Primates occasionally eat meat. Chimps have been known to hunt, though poorly. In spite of these facts, no primates pursue animals in the manner described above or use clubs to hunt. The creature observed in the 1860s seems to have done precisely that.

Hairy hominids often throw rocks at witnesses. Primates sometimes throw rocks too, but what if we're missing a crucial part of the rock-throwing phenomenon? Hairy hominids might do more than throw rocks. What if they use them as tools?

The evidence says they do.

Tools of the Trade

For years I had found large rocks lying around on my property, in places where none had been before. Although I noticed they hadn't eroded out of the ground, I still wrote them off as natural occurrences. Perhaps a rainstorm had washed them to that spot. The rainstorm explanation failed to account for the evidence, but my brain—as yet unaware of the strange creatures lurking in my woods—refused to consider other possibilities. I dumped the rocks into planters in my living room or dropped them outside the house after pondering them briefly.

Then the high strangeness began. UFOs, mysterious trespassers, out-of-place species—from the supposedly-extinct red wolf to glowing orbs, I started to see bizarre things in the woods (yes, "things" is exactly the right word for them!). For years, I had harbored an interest in the paranormal, a curiosity that had gradually intensified over the years, ever since I saw my first UFO at the age of fourteen. So, by the time I realized the sticks arranged in purposeful shapes derived from the actions of an intelligent being instead of dumb luck, my mind had reached the point where it could accept notions that once sounded wacky. I suppose they still do sound wacky. I've merely grown accustomed to wacky ideas proving true.

In May 2004, while still in Texas, I stumbled across a stone that I knew did not belong. The oblong stone, 3.5 inches long by 2.125 inches wide, lay in the middle of the main trail that bisects the woods behind my house. Although years of hard rains had eaten away trenches in the hillside along this trail, nothing larger than pebbles had eroded out of the ground. No large rocks occur in the vicinity. The stone I'd found was an inch and a half thick, smooth, its corners rounded. Water action must've worn down the stone, polishing its surface in the process. The nearest source of running water—a seasonal creek—lay half a mile away. The stone fit nicely in my hand. Even a larger hand would have no trouble grasping the stone, which seemed well-suited for use as a tool. I kept the stone, labeling it with the date and place of origin, yet I thought nothing more about the tool idea. One stone was interesting, but not enough to function as the basis for a theory.

I didn't have to wait long for the evidence to pile up on a shelf in the hallway of my house, in the form of rocks similar to the one I'd found in May. Dozens have turned up so far, 18 in Texas and the rest at my new home in Michigan. Most have similar features: a roughly triangular outline; a convex underside that fits snugly in your cupped palm; rounded edges; and a smoothed surface. They resemble the stones you'd find in a creek bottom. In fact, they *are* the stones you'd find in a creek bottom. My research showed the Texas stones originated in a creek bed half a mile from my house. No such stones resided on my property, except the ones someone dropped there. The stones I find in Michigan often exhibit the rounded corners and smoothness of creek stones, though many have sharp edges as if they've been chipped off a larger stone. No creek lies near my property. Nothing on my property could flake pieces off large rocks either.

One stone, which I discovered in Texas on August 26th of 2004, had been split in two. The halves lay 15 feet apart along the fence that delineated the southern property line. As usual, the rock's smoothness betrayed its identity as a creek stone. Most of the stones have retained their natural shape. The user, a.k.a. the stick-sign maker, chose them because they already possessed the appropriate shape for use as a tool.

Some of the stones have been reworked to obtain a better shape or a sharp edge. The stones which have not been reworked show signs of wear (use), including scuffs and scratches.

In Chapter Two, I mentioned the set of split stones that turned up in my field on July 30, 2005. This time, however, a smaller stone had been flaked off a much larger block. The smaller stone looks identical to stone tools attributed to *Homo habilis*. The larger stone bore indentations on two sides, as if the toolmaker had chipped flakes off both sides until he achieved the type of tool he wanted. The small flake either had been dropped by accident or dumped on purpose because the toolmaker had deemed it unsuitable. I could fit the small flake back onto the larger stone, in the precise spot where it had originated.

The majority of the stones I've found resemble a particular brand of stone tool familiar to anthropologists as "pebble tools." *Homo habilis*, the earliest scientifically-accepted members of the genus *Homo*, used pebble tools in their daily lives. Others of my stones bear an uncanny resemblance to scrapers used by *Homo habilis*. These scrapers would be flaked off a core stone, in the same manner employed by my mystery toolmaker. A few look more like the cleavers and handaxes fashioned by *Homo erectus*, the successor to *Homo habilis*. The toolkit used by *habilis* dates to about 1.7 million years ago. The *erectus* toolkit emerged around 1.5 million years ago.[42]

How did tools like those made by African hominids millions of years ago wind up in my backyard? The answer is simple. The stone tools got there because a modern-day descendant of the ancient hominids lives in the woods behind my house. We call the species Bigfoot.

The hairy hominids could well have kept the primitive, unshaped tools in their kit long after they developed more sophisticated tools. Even today, some humans use crude stone tools. The most sophisticated tools evidenced by the paleoanthropological record stem from the activities of modern humans. The double-edged blades, arrowheads, and spears appear at sites where humans lived during the last ice age. In comparison with those tools, the scrapers and handaxes seem downright archaic.

Often I discover the stones in my horse pasture, near the border with the woods. My horses have carved out freeways through the woods, paths they use so often their hooves have worn the ground down to the bare soil in a two-foot-wide, meandering line. No rocks may show up for days at a time. When I do find the rocks, however, they turn up on days when the horses act leery of the woods. They wait for me to embark on

The hammerstone found on my property, composed of a stone type not found on my land; the chip, which had been part of the larger stone, was found nearby and still fit onto the hammerstone (Michigan)—July 30, 2005.

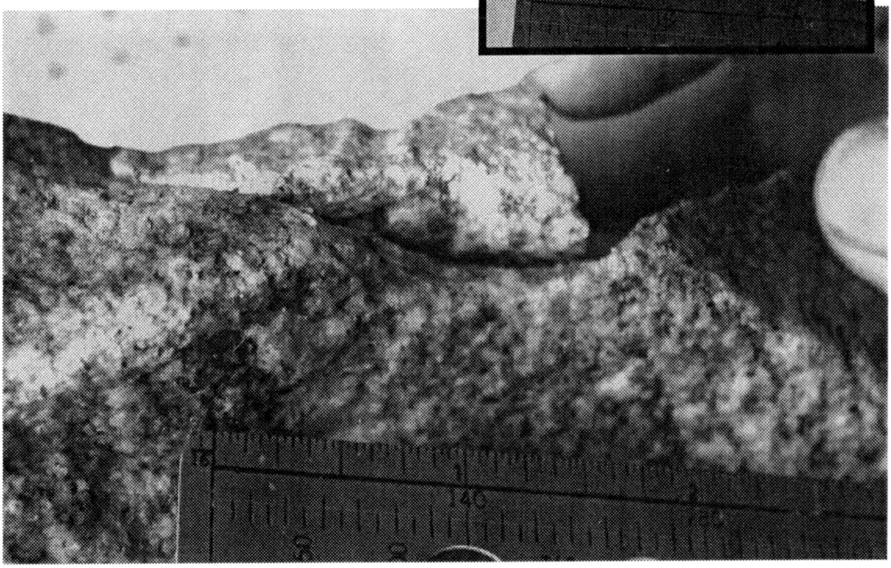

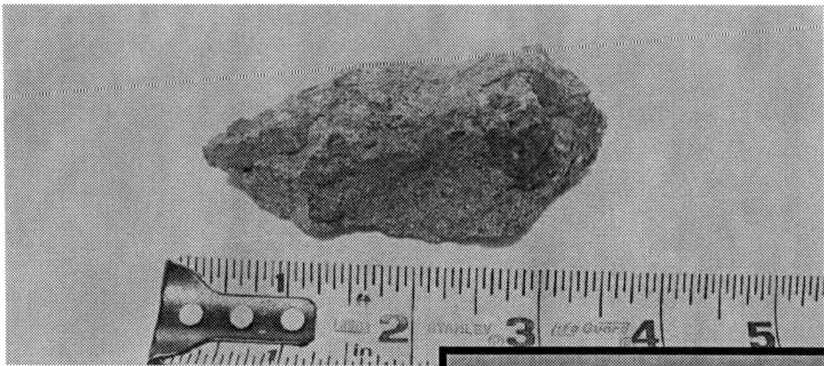

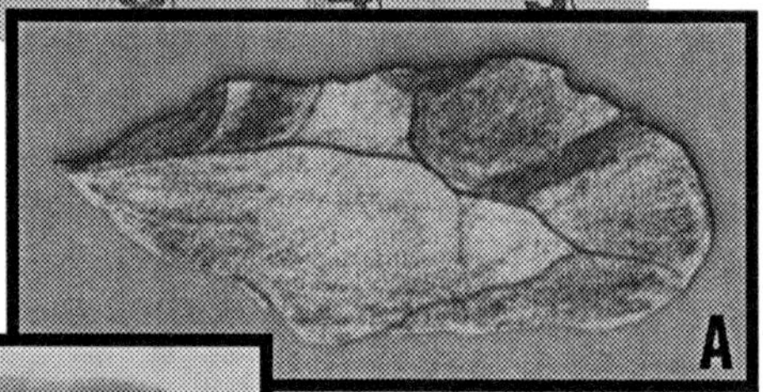

The stone in the top photo appeared on my property one day. The stone fits nicely in a hand (inset B) and has a sharpened point, with flakes chipped off the rest to shape it. The stone bears a striking resemblance to an Acheulian handaxe (inset A), a tool from the kit used by *Homo erectus*.

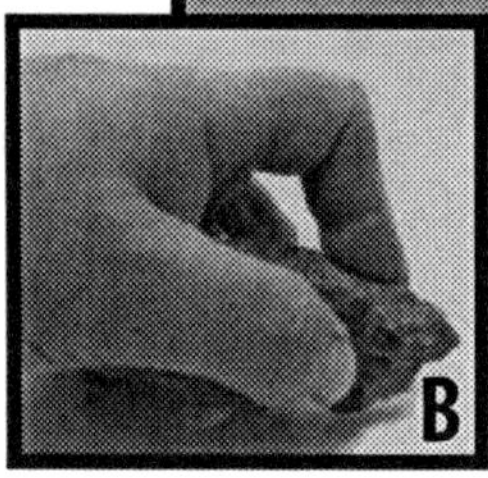

The stone in the bottom photo displays another shaped stone, though this stone has retained much of its smoothness. Like the first stone, this one resembles a type of ancient stone tool. However, in this case the stone looks like a pebble tool (inset C) from the kit used by *Australopithecus*, a predecessor of *Homo erectus*. The second stone also fits nicely in a hand (inset D).

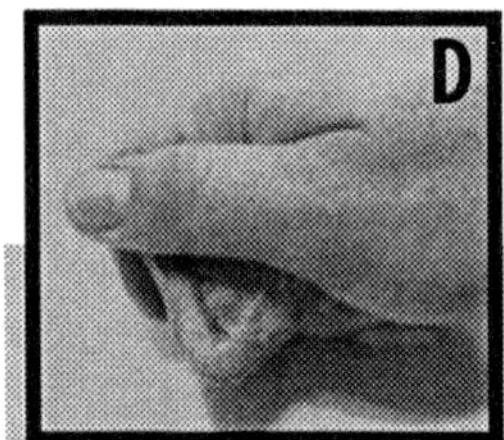

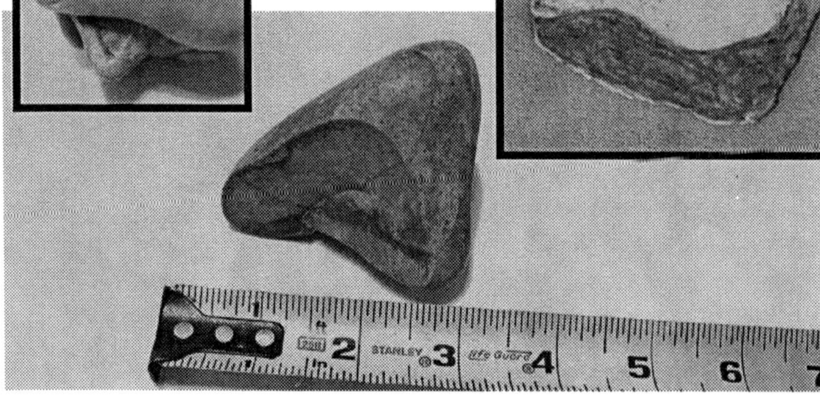

One of many tool-like stones I found on my property, both in Texas and Michigan. This stone, from Texas, had been split in two. The halves lay 15 feet apart along the fence line of my property. The stones appeared there overnight, and their smoothness evidences their origin in a creek bed a half mile away.

Top photo: The two halves, showing the seam where they fit together (Texas)—August 26, 2004.

Bottom photo: The two halves rejoined, showing where a flake had been chipped off one end to further shape the tool (Texas)—August 26, 2004.

A lineup of pebble tools I found on property over the years. Note the consistency in shape, despite the differences in size, as if someone chose them specifically for their shape. Ancient hominids also chose rocks of the right shape to use as pebble tools.

my morning walk in the woods before they will venture into the forest themselves. The rocks that I find lying in the field, on top of green grass, resemble pebble tools. They appear overnight near an area where the horses like to loiter.

I suspect the owner of the rocks throws them at the horses. I doubt he intends to harm the horses, or even to hit them, but lobs them as a game. Perhaps the rock thrower likes to see if he can spook the horses into running away. When I find the rocks in the field, I never find injuries on the horses. Either the rock thrower is an unbelievably bad shot, or he has no intention of striking the horses. The footprint I found in the swamp nearby was human-size, which I surmise indicates a juvenile. Human kids find enjoyment in odd activities, like squeezing frogs until they pee. Bigfoot kids may seek amusement in activities that we find equally silly.

Take, for instance, mane braiding.

Hair of the Horse

Another mystery emerged on May 8, 2005. One of my horses, a thoroughbred gelding, turned up with strange braids in his mane. The braids consist of locks of his mane interwoven and twisted together in a complicated pattern similar to a French braid, though vastly more intricate, with a tight knot at the end that is woven back into the braid. Studying the braid, I noted several extraordinary features. As someone who grew up having long hair, I know what normal braids look and feel like, from a simple three-strand braid to a French plait. The braids in my horse's mane differ from these. The technique used to create them is quite complex.

First, individual locks were twisted. Next, these twists were woven into a larger braid encompassing coiling, interlocking twists. Finally, the whole braid was finished off by tucking the end up inside the braid and tying a tight knot around it. These braids are difficult to untangle. In fact, I couldn't untie the strands; I had to cut them.

In Texas, I'd found similar braids in the thoroughbred's mane on numerous occasions. Back then, I was content to dismiss them as the handiwork of my neighbors. My thoroughbred is gregarious and loves children, so when I noticed several braids in his mane after the neighbors' grandchildren had visited, I assumed the children had woven the plaits.

But the braids reappeared several times, during intervals when no children had visited the rural neighborhood. After noting identical braids in the horse's mane here in Michigan, where I have no neighbors, I can no longer blame the plaiting on visiting children—at least not children of the human variety!

The mane-braiding echoes the legend from the island of Islay, in which the giant's daughters wore necklaces composed of twisted horse hair. Residents of Central and South America believe hairy hominids of that region braid the manes of horses.[22] While horses generally fear hairy hominids, they can become acclimated to any creature that at first frightens them. In Texas, my horses became accustomed to the wolves and cougars. Here in Michigan, they seem to have adjusted to the presence of hairy hominids, at least somewhat, as they now express a guarded interest in the areas where I've found stick signs, stone tools, and footprints. Lately, they have even behaved as if they enjoy interacting with the hairy hominids. One morning, my quarter horse mare showed up with a braid in her mane. Usually aloof, she acted excited and cheerful that morning, as if she'd had a grand time overnight.

If a young Bigfoot lurked around the horse pasture night after night, approaching the thoroughbred in stages, the creature could acclimate the horse to his presence. The delicate work required to weave the braids implies small fingers—such as those of a child, or a juvenile hairy hominid.

In Texas I had neighbors on whom to blame the braiding. Here in Michigan, my nearest year-round neighbor lives over 2 miles away, with acres of swamp between their house and my property. Whoever braids the horses' manes leaves no shoe tracks. Between May 2005 and January 2006, the horses have had their manes braided 22 times. The braid-maker

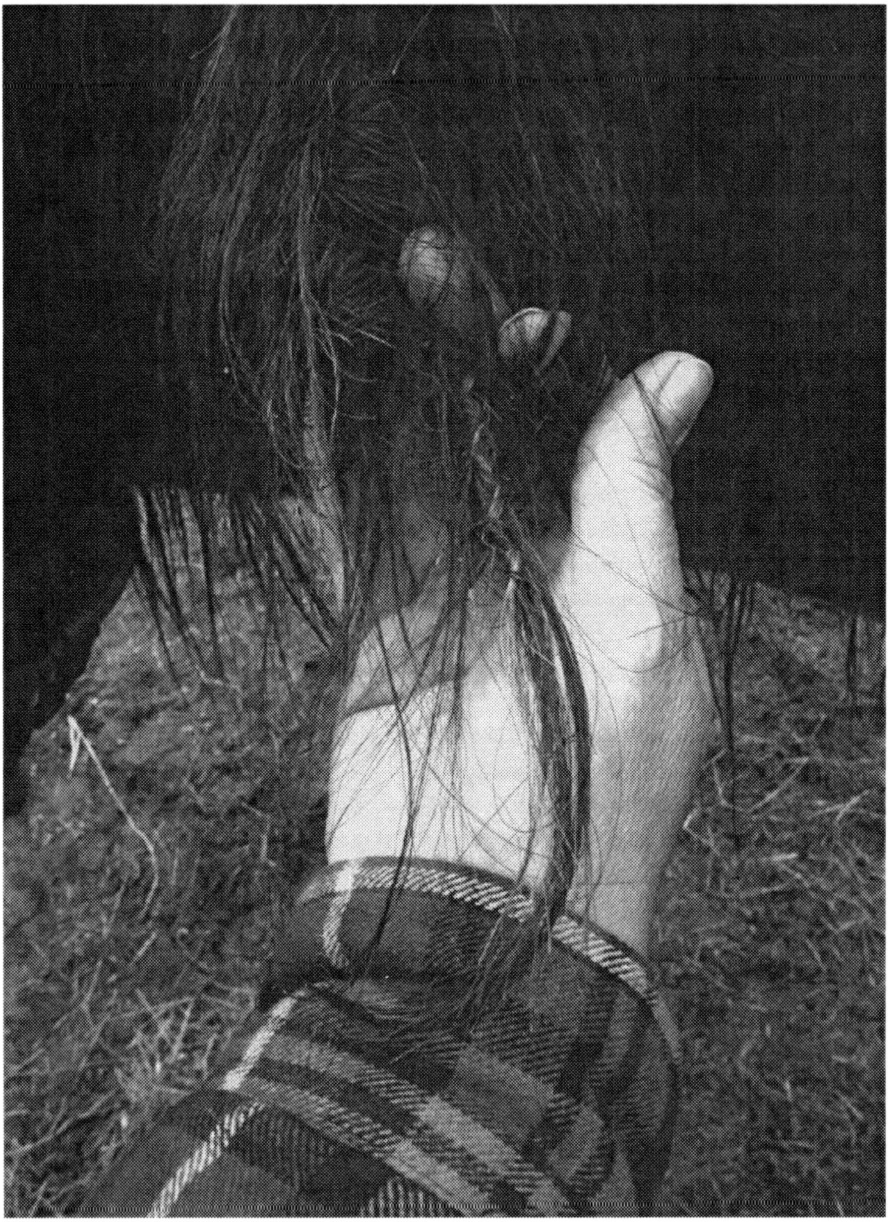

My thoroughbred's mane was braided many times, including this instance when an intricately woven braid appeared in his mane overnight (Michigan)—May 11, 2005.

With the nearest neighbor over 2 miles away, acres of swamp between them and my property, and no shoe tracks to indicate a human presence, I soon ruled out humans as the culprits. Besides, the braids are far more complex than any braid a human has invented. To create such a braid must take time and patience.

One the same morning, both my thoroughbred and my Tennessee walker geldings turned up with intricate braids in their manes (Michigan)—December 31, 2005.

The photo at left shows the Tennessee walker's braid. Individual hairs are twisted into one braid-like strand, then woven together in a tapestry unlike any human braid. The result is difficult to untie and often must be cut out of the horse's mane.

The top photo shows the thoroughbred's braid, which involved two thick twists joined by a knot.

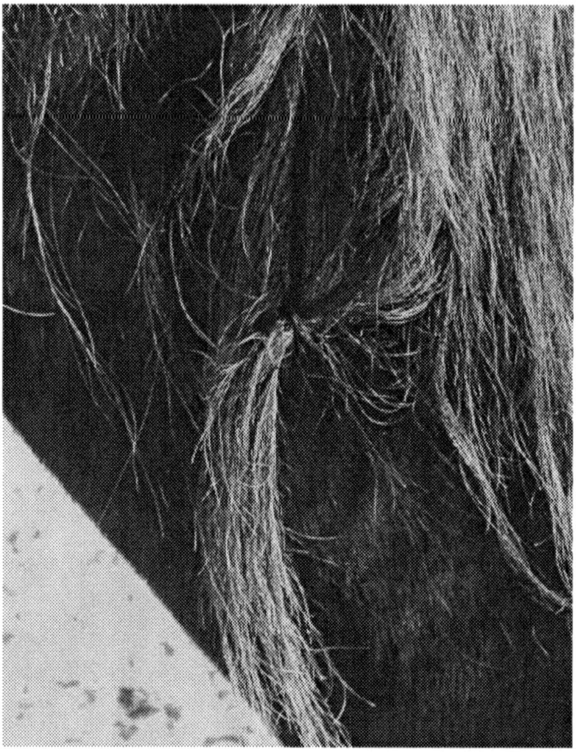

These two photos show different types of braids. Both appeared in my Tennessee walker's mane.

Top photo: A series of twisted loops deftly woven together into a knot that refused to come undone (Michigan)—September 20, 2005.

Bottom photo: Two tight braids knotted loosely into one strand (Michigan)—October 25, 2005.

seems to prefer nights when a snowfall is imminent, to cover his tracks with a few inches of powder. With surprising regularity, the braids show up the morning after a night-time snowfall. Since most animals can sense a change in weather, suggesting that hairy hominids know when a snowfall might occur hardly stretches the bounds of plausibility.

Here's another point to consider: horses seem to attract hairy hominids. Remember Roger Patterson and Bob Gimlin, the who captured a female hairy hominid on film back in 1967? Both Patterson and Gimlin were on horseback when they spotted the creature. Perhaps the creature had been attracted by the presence of horses. The hairy hominids in Texas and Michigan certainly enjoyed playing with the horses' manes, and placed their stick signs on trails frequented by the horses. The Michigan hominids in particular seem attracted by the horses. I believe owning horses may prove to be key to initiating contact with hairy hominids. While I have yet to see a hairy hominid in the flesh, I've heard their vocalizations and have benefited from years worth of interaction with them via the stick signs. If I'd not had horses, I doubt the interaction would've reached such a personal and playful level.

I mulled over other possible sources of the braiding. Wind might've tangled his mane, twisting the locks. But wind can neither weave individual locks into a braid nor loop the ends of the locks back into the braid in a complex terminal knot. The horse might also have gotten his mane entangled in a branch or briars. But, as with wind, neither branches nor briars can weave such an intricate braid into a horse's mane.

The perpetrator who braids my horse's mane, throws rocks, and shapes stone tools from core blocks must possess a keen intelligence and a keener curiosity. Does evidence from other regions back up this statement?

Yes, it does.

Brainy Bigfoot

Their curiosity shows up in how they act toward humans. Hairy hominids often watch humans from a distance. They peek out from behind trees or brush, or peer through windows of cars or homes at startled witnesses. In the 1930s near Priest Lake, Idaho, several witnesses found themselves under surveillance by a hairy hominid. The creature knelt on a bench outside the window of a ranch house to gaze in at the humans. Upon attracting their attention, the hominid strolled away.[33] A Portland, Oregon, newspaper reported the encounter between a lone fisherman and a hairy hominid that occurred in the summer of 1953, when the creature observed the fisherman from behind some brush.[7] In 1966, a hairy hominid peered in a window at a couple in Lower Bank, New Jersey; when the couple left food out for the creature, he took it.[33] A volume of experiences related by 19th Century pioneers, titled *Told By The Pioneers*, includes a Bigfoot sighting that happened in the 1850s around Mt. St. Helens, Washington. The daughter of witness Rocque Ducheney described how a hairy hominid had beckoned to her father, who fled in the opposite direction.[7]

In several instances, the creatures paced vehicles and in some cases the witnesses estimated the creature's speeds at up to 80 miles per hour. In at least one case the hairy hominid nicknamed MoMo, short for Missouri Monster, attempted to open the doors of a car in which the frightened witnesses had taken shelter[33]. When they do chase humans, the witness usually escapes without much effort. In at least one case, when the witness fell down during the chase, the hairy hominid paused to let him regain his footing.[7] Since a creature as large as Bigfoot could've outpaced the witness, the conclusion seems inevitable. They're as interested in us as we are in them.

I've experienced hairy hominid curiosity firsthand. Twice I have seen a large, brown animal watching me from within the murky veil of the woods. The first time, I thought at first I had spotted my thoroughbred—until I noticed all three horses grazing on the opposite side of the field. Since my thoroughbred stands 17 hands high, whatever I saw must've been quite big. On another day, while walking in the woods, I saw something large and brown keeping tabs on me from deeper in the woods. When I looked at the watcher, it moved off out of sight.

Eyewitnesses often describe a sensation of being watched, before they see the hairy hominid. Many times, they never see the creature, but feel something observing their actions and hear noises in the brush. I've felt this sensation too, which I used to dismiss entirely. When I started hearing noises in the woods, sounds which I could not tie to any normal animal, I wondered but still rejected the stranger verdict. The whooping, thumping, crashing, and growling noises could have a prosaic explanation. Porcupines can make an assortment of noises that includes growls, pileated woodpeckers can bang on trees as if they have beaks made of steel, and owls and loons can vocalize calls that snake a shiver up your spine. Yet none of the mundane explanations account for the fear my dogs have evinced during and after the noises, or the wary curiosity my horses exhibit toward the wooded precinct where I've seen large shapes and heard odd sounds.

The morning that a stick sign appeared right outside my house, inside the horse pen in front of the barn, my dogs hid upstairs the entire day. Despite the sweltering heat of a 90° day, which turned the upstairs into a sauna, the dogs refused to stay downstairs. On any other day, especially a hot one, the dogs lounge in the living room where a fan ventilates the room. On another night, a horrendous yowling woke me at 1:00. I soon recognized the sound as a cat growling-yowling, though I'd rarely heard a cat cry with such volume. The cat seemed to be moving toward the pole barn from the woods on the far side of the barn. By the time I got outside, the cat had quieted. I finally located my cat in the pole barn. Knowing now that I had come outside, he relaxed and followed me inside. He was still frightened, however, as I carried him upstairs into my bedroom. He lay beside my pillow the rest of the night.

While far from conclusive, these incidents hint that something scary enough to panic the heartiest animals lurks in the forest around my home. Coyotes and wolves don't terrify dogs or cats. A bear might, but bears don't stand 17 hands high as they stroll off into the woods. Michigan hosts black bears, not grizzlies, and I know what black bears look like—a few months ago, one crossed the road in front of my car. Whatever inhabits the forest is large, brown, and curious.

A few sightings mention the creatures wearing clothing, typically an animal skin covering the lower body. In the early 1960s, a witness spotted an adult hairy hominid, a female, holding the hand of her juvenile companion as they walked along a river bank. We can only wonder whether the youngster was her child or perhaps a younger sibling. Then on Mount Shasta, around 1900, a man was bitten by a snake and passed out, awaking later to find three hairy hominids had treated his wound. The creatures carried him to a tree in an area where he could attract the attention of the prospectors for whom he worked.[7] Behavior like this echoes human compassion.

If, as seems most probable, hairy hominids descended from the ancient hominids, learning about the behavior of those creatures might tell us more about modern hairy hominids. The fossil record offers little in the way of cultural or behavioral information on the ancient hominids. But archaeological evidence can illuminate their lives.

Creatures of Culture

Everyone learns at least the basics about hominids in school. All humans evolved from apelike creatures who miraculously learned to walk upright, abandoning the trees for the savannah. Over millions of years they progressed from three-foot, chimp-like beings into humans through a sequence of adaptations perpetuated by natural selection. Yet few of us learn about the other side of hominids.

For years, the prevailing image of ancient hominids—in particular Neanderthals—portrayed them as violent brutes with brains about as developed as squash. No wonder calling someone a "Neanderthal" became an insult! Yet recent discoveries have overturned the "killer ape" notion. In contrast, the ancient hominids seem just as capable of forethought and culture as modern humans. Consider for a moment the stone tools they created.

The earliest examples of stone tools date from 2.3 million years ago, the age of *Homo habilis*. The tools consist of stones reshaped by the user to form a simple blade. The user hits a larger stone with a smaller one to knock off pieces, called flakes, in a process known as "flaking." Until the 1990s, few anthropologists would've considered the idea that the earliest hominids used their intellect to craft more complicated tools, or that their flaking techniques exhibited planning, analysis, and precision. But the tools unearthed in Kenya during the 1990s illustrated just these traits. The toolmaker removed as many as 30 flakes from a single stone to achieve the perfect shape.

Examining other examples from the same region showed that the toolmakers worked according to a defined process, rather than randomly flaking off bits until they happened upon the right shape. Having discovered the ideal angle for striking the stone, the toolmakers also duplicated the angle each time, resulting in a remarkable similarity between the tools.[13] These "ape-men" clearly had the forethought and analytical skills to recognize the best angle, reproduce it with multiple stones, repeat the flaking process until they got it right, and teach their skills to their brethren.

A question arises here. Did the early hominids have intellects equal to modern primates, or had they advanced beyond that level? The chimp-like physique of *Australopithecus*, and the continuation of apelike features in varying degrees in subsequent hominids, may lead us to believe the early hominids exhibited no greater mental prowess than a chimp or gorilla—intelligent animals, for sure, though far removed from humans. One group of researchers undertook an experiment to determine whether the tools attributed to *habilis* required human- or primate-level intellect. They showed a bonobo the utility of stone tools for cutting open boxes of treats, then demonstrated a typical method for making the tools. The bonobo started to make tools on his own, and even invented his own method of hurling the stone into a hard surface to break off flakes, but he has yet to reach the level of expertise seen in the earliest stone tools from the archaeological record.[56] The evidence illustrates the differences between ancient hominids and primates. While they may look similar in many respects, their cerebral capabilities divide them like a Grand Canyon of brainpower.

Further evidence emphasizes the intellectual skill of early hominids. In the Lake Natron region of Tanzania, in Africa, analysis of animal bones has revealed startling facts about the early hominids. The age of sites that contained bones and tools, estimated at 1.4-1.6 million years, suggests that *Homo habilis,* whose remains have cropped up in the area, laid claim to the tools found there. The animal bones featured cut marks and tooth marks, a sign that hominids had defleshed the carcasses. The popular theory has hominids scavenging whatever carcasses they stumbled over in their wanderings, presumably after predators and other scavengers had taken their turns. But at one site, hominids traveled some distance to find a meal, which they then brought back to their home territory—as evidenced by the animal bones found there.[45]

Hominids must've searched for carcasses, or even hunted the animals themselves, then transported the carcasses to another location where the flesh was removed using stone tools. The hominids didn't live where they defleshed their carcasses. Instead, they seem to have chosen one site as their butcher shop, and another for their living quarters.

Then we have *Homo heidelbergensis,* an *erectus* successor who migrated into northern Europe around 600,000 years ago. At Boxgrove in England, scientists have uncovered a hammer deftly sculpted from the antler of a giant deer called *Megaloceros dawkinsi,* a species that surrendered to extinction 500,000 years ago. Fragments of flint on the hammer's surface demonstrate that *heidelbergensis* used such hammers to shape stone tools.[42] To create a hammer for repeated uses implies *heidelbergensis* had the ability to plan ahead, as well as to recognize the value of making one tool to aid in the manufacture of others.

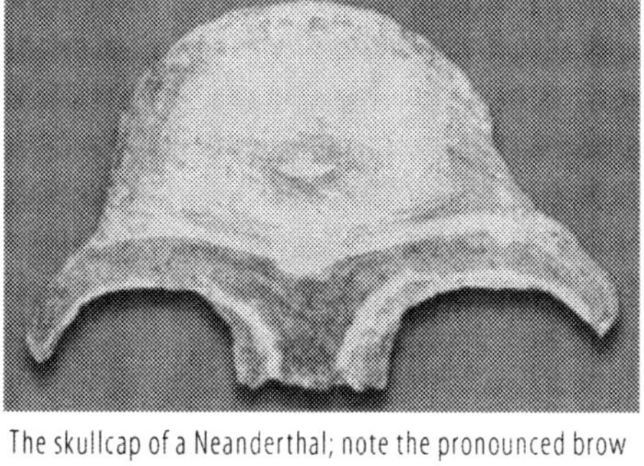

The skullcap of a Neanderthal; note the pronounced brow ridge, similar to that of a hairy hominid.

By 400,000 years ago, *heidelbergensis* had begun to

craft wood spears, the ends of which these ancient experts split open to form a snug slot for the spearhead. They apparently realized which trees offered the hardest wood and therefore made the best weapons. Interestingly, *Homo heidelbergensis*—once known as archaic *sapiens*—sported shins reinforced by twice as much bone tissue as in modern humans and marked with ridges etched by huge, powerful muscles.[42] Both features indicate a brawny hominid who shared more in common, physically, with modern hairy hominids than modern humans.

Neanderthals, despite early misconceptions, behaved as anything but dim-witted cavemen. On average, Neanderthal brains weigh in heftier than human brains, though argument about how that pertains to brainpower remain unresolved.[63] Neanderthals did develop the most sophisticated toolkit of any hominid, excluding humans (who may not be true hominids anyway). Kebara Cave in Israel, a source of plentiful data about Neanderthals, contained over 25,000 stone tools—a number of which the "caveman" residents seem to have incorporated into spears and/or lances.[42] Traditionally, spears and other compound tools demarcate the imaginary line between the ancient hominids and modern humans, into whose dominion such items fall. Kebara Cave challenges the notion that the ancient hominids lacked the brainpower to craft advanced weapons and tools.

Along with hewing advanced tools out of stone, Neanderthals may have possessed a softer side. Objects with holes punched through them found at Neanderthal sites may have served as pendants for necklaces. Remains of elderly, infirm Neanderthals discovered at Shanidar in Irag hint that members of the species cared for their elders.[63] When their kinsmen died, they also buried them in a purposeful manner that suggests ritual and perhaps a concept of the afterlife.

Neanderthal Funeral

A number of sites preserve Neanderthals burial—individuals purposely interred, rather than covered by sediment from flood waters or crushed under falling stones. Several examples have surfaced in Israel. In a cave called Amud, Israeli paleoanthropologists unearthed the 50,000-year-old remains of a well-preserved infant laid out with its arms at its sides, the jawbone of a red deer positioned over its hips, and the whole covered over with dirt.[63] Nobody would think twice about such a discovery if not for one fact—the child was clearly not human. Paleoanthropologists have described the remains as "alien,"[63] an odd term to use for a Neanderthal child, since Neanderthals are considered cousins of humanity.

Paleoanthropologists exposed another Neanderthal burial at Israel's Kebara Cave, where a male lay in a pit widely believed to have been purposely dug and filled in after interment about 60,000 years ago; a similar scenario unfolded at La Chapelle-aux-Saints in France, in which fellow Neanderthals interred a male inside a simple, rectangular pit with sides too regular to have eroded out by natural means. Two additional sites in France impart clues about Neanderthal beliefs: at Le Moustier, Neanderthals buried a teenage boy with care atop a bed of flint flakes, settling his head on one arm; La Ferrassie preserves a cemetery where two adults and four children were laid to rest. [12]

Down in Australia, Kow Swamp furnishes additional evidence of purposeful treatment of the dead. In the last chapter, I discussed the Kow Swamp "people" and their resemblance to Neanderthals and *Homo erectus*, despite their rather arbitrary classification as *Homo sapiens*. These seemingly primitive beings nonetheless cared for their dead in a manner that echoes the Neanderthal burials of Europe and the Middle East. About 10,000 years ago, these Neanderthal-like "people" had partially cremated their dead, then sprinkled red ochre over the remains.[63] For humans to engage in ritual burial 10,000 years ago triggers no mental discomfort, but for a people who resemble—and therefore likely are—ancient hominids to prepare the dead for burial paints nonhuman bipeds in a different shade altogether.

Far From Primitive

They buried their dead. So what? If the ancient hominids had human-level culture, we should've found other signs of it.

Controversial evidence from Indonesia suggests ancient hominids sailed the ocean. Stone tools found on the island of Flores were at first attributed to *Homo erectus*. Since the discovery of *Homo floresiensis*, the dwarf hominid, anthropologists have argued about whether the tools belong to *erectus* or *floresiensis*. The only remains of *floresiensis* found so far date to between 18,000 and 12,000 years ago. At least one site that yielded stone tools dates to around 800,000 years ago. Yet no remains of *Homo erectus* have turned up on the island, which leaves us with *floresiensis* as the culprit. Whichever species created the tools, their presence on Flores raises a question that as yet has no answer.

How did the toolmakers get to Flores 800,000 years ago? Although land bridges connected other archipelagos at the time, nobody has found evidence of a land bridge joining Flores to mainland Asia. Hominids had one way of reaching the island. They must've sailed there on boats.

Detractors of the idea have agreed that hominids must've immigrated to the island via sea routes, but they insist these "primitive" people clung to flotsam, drifting across the ocean to land on Flores by sheer chance. I'd like to see the cynics try that one. Why should creatures intelligent enough to make and use stone tools decide to leap onto a hunk of driftwood? What possessed them to cling to debris for days or weeks until they washed up on Flores?

The hominids of Flores used their brains to survive on an island full of giant lizards, giant rats, and carnivores. Why should we refuse to believe they could build a boat? Curiosity drives humans to build craft and embark on journeys into regions as unknown and alien as outer space. The ancient hominids might've felt a similar urge to explore, a desire that brought them to Flores.

Or, perhaps, someone else took them to Flores.

Giant, But No Giganto

The "Giganto theory" crumbles under the pressure of scrutiny. The creatures we call Bigfoot, or thousands of other regional names used around the world, resemble not giant apes, but hominids. Most Bigfoot researchers deny that hairy hominids use tools, for to admit they do would crush the Giganto theory into bits too small for anyone to reconstruct it.

Meanwhile, the disappearance of the Neanderthals has baffled paleoanthropologists for decades. The archaeological record suggests that Neanderthals possessed the smarts and the social bonds not only to survive, but to challenge humans. Why then should Neanderthals have lost out, vanishing into extinction? The oft-ignored evidence concerning hairy hominids shows that the Neanderthals did not disappear or become extinct. They and their hominid kinsmen live on today in the guise of Bigfoot and other hairy hominids.

Apes sometimes make simple tools, although your idea of what constitutes "making a tool" colors any discussion of primate tool use. Chimps use the sticks at hand to fish for termites, or rocks at hand to crack the shells of fruits.[24] They can pick grass and cut it to length to perform similar tasks or sharpen the end of a stick. Sometimes they appear to choose a tool ahead of time, before they reach the termite mound.[13] Does this behavior truly represent tool-making? It reveals

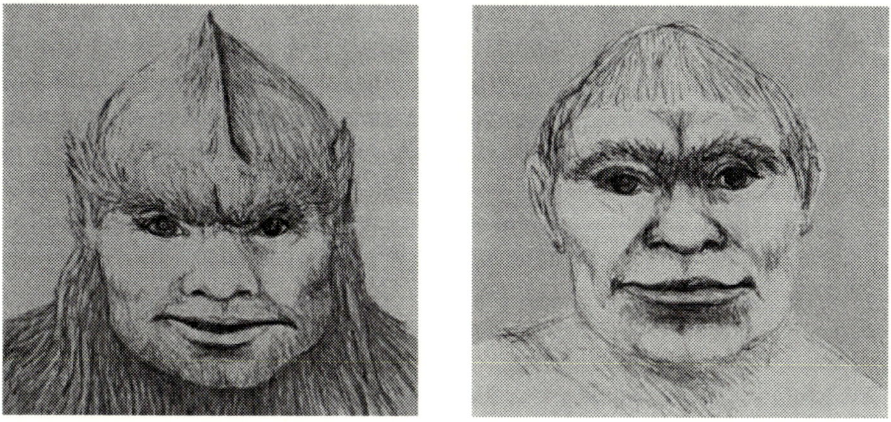

Artist's renditions of a typical hairy hominid (based on witness descriptions and photogrpahic evidence) and *Homo erectus* (based on skeletal evidence from Europe, Indonesia, and Africa)—note the similarities in facial features and skull shape.

their intelligence, for sure. Even if we call the sharpened sticks and cut grass tools, the chimps needed little forethought and no societal organization to manufacture those tools. They spotted a ready-made, or readily altered, implement and exploited it.

Some researchers have attempted to duplicate the "tool-making" of chimps, only to discover no surefire method exists for determining ahead of time which type of tool will work best for retrieving termites, and that chimp tool-making appears to derive purely from trial-and-error.[35]

When confronted with a water source out of their reach, say in a crevice, chimps will sop up the water with crushed leaves in order to drink it. They sometimes clean themselves with leaves too.[14] If we consider these situations instances of tool use, then we should also conclude that horses use tools. Horses will search out a good sapling, position their bellies over it, and rock back and forth to scratch themselves.

Most of the tool use by primates has occurred in captivity under the supervision, and with the encouragement, of humans. I'm not denigrating the intelligence of primates. I believe animals possess enormous intelligence and ingenuity. But the innovation and forethought required to craft and utilize complex tools, form societies, and develop systems of belief seems to occur in just two types of animal: humans and hairy hominids.

Given the evidence that modern humans appeared deeper in the shadows of prehistory than we thought, humans may well have shown the early hominids how to flake bladed tools off a core, or any of the other techniques which advanced stone tools beyond the unshaped pebble variety. But fashioning a stone tool requires forethought, precision, and determination. For tool-making to catch on, the original innovators need to teach their offspring the methods they developed for creating tools. Passing on knowledge demands culture. A diffuse group of hominids with no social bonds could never have achieved the synergy necessary to cultivate a toolkit. The tool-makers had to perfect their craft, then teach it to the next generation. Innate knowledge simply cannot account for the sophistication and continuity of the toolkits used by ancient hominids—or the hairy hominids of today.

The tool use by primates demonstrates that their cousins, the hairy hominids, had the right background to become tool users and toolmakers. Yet they seem to have undergone an astounding transformation from ape to ape-man. What triggered that transformation?

Another, related phenomenon may provide clues.

5
Orbs in the Night

Just as the stick-sign enigma endures, so does the mystery of the strange lights observed and captured in photographs that accompany both stick signs and hairy hominids. Thus far we've examined copious evidence that hairy hominids have skulked among and around us for as long as we've inhabited the planet, perhaps longer. For just as long, we've witnessed strange aerial phenomena that defy explanation and strain the limits we strap onto our reality. Flying chariots, flying hats, god's boats, foo fighters, flying saucers—throughout the ages, humans have attached many names to these phenomena. Since the mid-20th Century, we have called them unidentified flying objects—UFOs.

What do UFOs have to do with Bigfoot?

Nothing, cryptozoologists would say. Science, of course, scoffs at the notion that either hairy hominids or UFOs exist, dismissing both as misidentifications, figments of imagination, or outright hoaxes. Most Bigfoot researchers refuse to examine any evidence that links the two phenomena.

To do so, they say, would diminish the chances of scientists ever taking the Bigfoot phenomenon seriously. If we want science to accept the reality of hairy hominids, we must ignore the paranormal aspects of the phenomenon. Hairy hominids are giant apes, creatures of flesh and blood. The paranormal has no bearing on them.

Here I must pause to clarify a point that has caused much confusion, and elicited the ire of many Bigfooters. Those who mock UFOs, and think the paranormal has no place in Bigfoot research, insist that to connect hairy hominids with UFOs means to turn them into metaphysical beings. For these people, paranormal equals nonphysical. But the term paranormal simply means "outside the normal experience." The term refers to any phenomenon that remains unrecognized or unexplained by science. Hairy hominids, whether you connect them with UFOs or not, *are* paranormal creatures. Yet most Bigfooters deride the UFO connection as paranormal—i.e. silly, crazy, not "flesh and blood."

I find it interesting that Bigfoot researchers employ the same tactics in denying the UFO connection that so-called skeptics use in debunking the Bigfoot phenomenon. Ridicule it, misrepresent it, and pray it sinks into oblivion, never having touched you. You're entitled to doubt the existence of UFOs, to ignore the evidence of a hairy hominid-UFO connection, to believe the creatures are nothing more than giant apes. But ridiculing anyone who disagrees with you serves no useful purpose—unless your purpose is to silence the other camp.

Yes, hairy hominids are flesh and blood creatures. Like deer, bears, and human beings, they exist as physical animals. Yet a significant, and startling, mass of evidence reveals a link between UFOs, the beings who pilot them, and hairy hominids.

As for psychic Sasquatches and the ghosts of dead Bigfoot, these things may exist too. I see no reason to dismiss them out of hand, though I have seen no evidence for either that convinced me. I keep my mind open to the possibilities, for possibilities are boundless.

The same applies to the Giganto theory. Its proponents may prove right one day, although the evidence suggests their theory lacks the substance to persevere. Hairy hominids may be giant apes with no connection to UFOs.

The opposite may also prove true. At this point, no one theory can be proven beyond doubt. None of us can do more than examine the evidence and reach our own conclusions.

This chapter focuses on the evidence for a connection between UFOs and hairy hominids. I won't try to prove UFOs are real or from what galaxy they hail. Countless books have presented the evidence for the reality of UFOs. Any reader interested in learning more about the physical and anecdotal evidence for the existence of UFOs can browse Amazon.com or the local library for a plethora of choices.

Instead, I will focus on providing a historical perspective for the connection between UFOs and hairy hominids. Then I'll wrap up the chapter with an account of the high strangeness in my own life. Are these events connected with hairy hominids? I believe they are. Am I right?

Decide for yourself.

Roots of an Enigma

Fairies and Dwarves

When people see hairy hominids, their minds process the event in a manner their life experiences, religious beliefs, and upbringing can handle. Within this framework of acceptable conclusions, people find a label for the creatures they've seen. The first Europeans to explore North America spoke of "wildmen" or "hairy Indians." UFOs used to get tagged as angels, demons, fairies, or spirits. More recently, some scientists have tried to dismiss the entire UFO phenomenon as "earth lights"—energy released by seismic activity that appears in the form of lights. Like the witnesses at Fatima, Portugal, who believed they'd seen the Virgin Mary, the scientists who concocted the earth lights explanation have labeled the phenomenon based on what fits inside their mental scaffolding.

How did people in ages past react to hairy hominids and UFOs, seen together? A large body of folklore, anecdotes, and eyewitness testimony reveals the answer.

In Brittany, folks knew about the *fions*, hairy dwarves, dark in color, who came out at night with the fairies. They had black eyes and carried small torches.[65] The fairies themselves were associated with UFOs called "fairy lights," which witnesses often saw dancing above the hills. Early in the 20th Century, a young girl watched fairy lights—white lights—soar in formation from the sky above the hill fort of Crillaun toward another hill fort. She and her friends had seen the lights on other occasions too.[8] Legends of hairy dwarves persisted in the Poitou region of France up until the late 19th Century. Here, the creatures—called *farfadets*—were described as hairy dwarves, dark in complexion, who slept in caves during the day, emerging at night to taunt farm folk. In the 1850s, several women watched in fear as a group of *farfadets* pulled a "chariot" uphill toward them. The "chariot" then flew up into the sky and vanished from sight![65]

Earlier I mentioned the *ikals*, hairy dwarves reported by residents of Mexico since ancient times. The *ikals* are about three feet tall, dark like the *fions* and *farfadets*, and also quite fierce. Witnesses have seen them in conjunction with glowing spheres. The *ikals* live in caves like the *farfadets*. Ancient tales say the *ikals* could fly, while modern reports state that the *ikals* wear rocket packs which enable them to fly.[65] Why would the ancient stories not mention rocket packs? The ancient Mexicans knew nothing about rocket packs—or rocket ships, or bottle rockets; therefore, they had no frame of reference for describing what they saw. It makes sense that ancient stories would say only that the *ikals* fly, without mentioning rocket packs. Perhaps the legends of the *ikals* can help explain where the notion of winged fairies originated.

Clear across the Atlantic, in France, children playing outside one evening in 1954 spotted an orb-shaped craft about eight feet in diameter, supported by three legs, which had landed near a cemetery. A small being exited the craft. He held a bright light in one hand, which obscured much of his features, but the children could tell he had a hairy face. A loose garment covered his body. The creature spoke to the children, then shut off his light. Frightened, the children ran away, though they did look back to see a bright light zipping across the sky.[65]

"Tall" Tales

As the 20th Century ticked onward, the hairy hominids sighted in conjunction with UFOs shifted toward the taller end of the spectrum. The days of hairy dwarves had given way to the era of the Bigfoot.

In August 1955, Mrs. Darwin Johnson decided to take a swim in the Ohio River near Dogtown, Indiana. Fifteen feet from shore, a hairy, clawed hand seized her leg. As the hand towed her under, she managed to break free of the creature's grip. An account of the incident in the local newspaper spurred several people to report what they'd seen at the same time, in the same area—a flying saucer.[7] Since no one saw the creature itself, debunkers could argue the woman mistook a piece of driftwood for a clawed hand. However, Mrs. Johnson later found the imprint of a hand on her leg. The claws on the creature's hand could easily have been long fingernails sharpened into talons. Any creature as close to humans as a hairy hominid must have fingernails. Sharpening them could have uses for a woodland creature.

Nearly 20 years later, in August of 1972, the Rogers family of Roachdale, Indiana, experienced a number of nocturnal visits from a hairy hominid. The creature would thump on the outside of the house. When a family member would run outside to catch the beast in the act, the hairy hominid would flee into a nearby cornfield. About an hour before the first visitation, neighbors had spotted a glowing object hovering over the same cornfield. The object seemed to explode, without leaving debris or any sort of evidence of a crash.[7]

A man fishing near Galveston, Indiana, in October 1973 witnessed a similar event. First, he noticed a sandy-haired hominid observing him from about 20 feet away. The witness decided to try speaking to the creature, which seemed to frighten it, for it took off into the trees. Soon after, the fisherman felt something tap his shoulder. He turned to see the Bigfoot. In that playful manner common to Bigfoot sightings, the creature then galloped off across a road—with the witness in hot pursuit. Before the fisherman could catch up to the creature, the hominid vaulted over a ditch and ducked into the trees. At precisely that moment, a golden light ascended from the same area in the trees. The object whizzed away, out of sight. The witness revisited the area two days later, this time with his fiancée and friends. A UFO tailed them most of the way and, though they initially spotted the 8- or 9-foot creature alongside the road, they lost track of it when an approaching vehicle forced them to drive away from the area for a few minutes.[7]

In the eastern US, 1973 proved to be a banner year for Bigfoot-UFO sightings. Just a few weeks before the fisherman's encounter—on September 27, in a rural district of Westmoreland County, Pennsylvania—two girls spied a 7- to 8-foot, white hairy hominid whose eyes glowed red. The creature, who lingered in the trees near the girls, held a glowing orb in its hand. When the girls scurried home to tell the tale, their father immediately tromped out to the scene of the incident to investigate. Although he was gone for over an hour, afterward he denied having entered the woods at all. Stranger still, he ordered everyone else to stay out of the woods too. His only comment on the incident consisted of a warning that the whole thing was better forgotten.[7] Westmoreland County has seen more than its share of strange occurrences, including the 1965 crash of a UFO at Kecksburg.

The best-documented case involving both hairy hominids and UFOs occurred on October 25, 1973, in Fayette County, Pennsylvania. Fifteen spectators had witnessed a red UFO descend on a pasture. Three young men resolved to investigate, and set out for the pasture in which the UFO seemed to have landed. In the pasture, they found the object—a glowing, dome-shaped craft about 100 feet wide—which had turned from red to white. A strange cry originated from the vicinity of the object, as did a terrible smell (a common element in Bigfoot sightings). At the same time, the boys noticed a pair of 7- to 8-foot hairy hominids traipsing down the fence line toward them. The boys had guns with them, and one of them fired a tracer round over the creatures' heads. Just as one creature raised its hand, the UFO vanished. Subsequent shots failed to affect the creatures, who simply walked away.

The boys ran home. The police were called in, and a state trooper arrived to inspect the scene. In the area where the boys said the UFO had landed, the trooper found a circle of earth glowing so brightly it outshone his flashlight.[31]

A few months later, in February of 1974, a woman living in a rural part of Pennsylvania, near Uniontown, grabbed her shotgun and opened her front door to check out the noises emanating from the porch. She found herself confronted by a 7-foot hairy hominid. The creature threw up its hands. Fearing an attack, the woman shot the creature point blank, but the hominid vanished in a flash of light. Her son-in-law, who lived 100 feet away, had heard the gunshot. When he phoned to check on his mother-in-law, she related her experience. Apparently sensing her fear, the son-in-law grabbed his revolver and marched over to his mother-in-law's house. Before he could reach the house, however, four or five hairy hominids with eyes that glowed red intercepted him. Understandably scared, the man fired two rounds before taking off for his mother-in-law's house. Both witnesses observed a flashing red light that floated above the woods 500 feet away.[7]

So far, none of the sightings have mentioned the creatures entering or exiting a UFO or standing inside one. In 1974, William Bosak saw just that near his home at Frederic, Wisconsin. Driving home one night, he saw a disk-shaped UFO partly shrouded in mist. The craft sported a windshield of sorts, in the form of a curved window. Through the window, Bosak could see a hairy hominid. He studied the creature for several seconds, until fear overcame him and he sped away.[7]

A recent case occurred on July 4, 1996, near Fayetteville, Arkansas. A witness reported that he and several friends had left a holiday gathering to walk into the woods near his home. A bright light up ahead attracted their attention. Worried someone was shooting off fireworks inside the woods, they moved toward the light. Through a break in the trees, they saw three triangular UFOs hovering in the woods, each pointing beams of light downward from its three corners. Three creatures, whom the witnesses took for the pilots of the UFOs, stood outside the craft talking to each other.

The creatures were tall and covered with hair. Sniffing the air, the creatures seemed to sense they'd attracted an audience, though the witness felt sure the creatures hadn't seen him or his friends. The creatures retreated into their craft. The UFOs took off in a blinding flash of light, knocking the witnesses to the ground.[58]

A Patchwork of Evidence

Many more sightings involve locales where witnesses have separately seen UFOs and hairy hominids. Many Bigfoot researchers have rushed to the vicinity of a recent sighting only to encounter UFOs in lieu of the expected Bigfoot. Some people have reported seeing hairy hominids who wore glowing belts, or metallic belts with buttons on them. In his book *The Locals,* Thom Powell describes a number of cases that fall into the above categories, as do Janet and Colin Bord in *The Bigfoot Casebook* and *Alien Animals.*

I have described some Bigfoot-UFO cases in order to provide the appropriate backdrop for the experiences I'm about to recount. These occurrences happened to me personally, at the locations where I found footprints, stick signs, and other physical evidence. None of the stories I will relate involve hairy hominids in combination with UFOs. However, all the events detailed in this book—from the strange cries in the night to the footprints and stick signs—occurred in the same small areas as the UFO sightings. I find it difficult to believe the phenomena are not connected, especially considering the anecdotal evidence pointing to a relationship between these bizarre phenomena.

Tex Files

As an eyewitness myself, I know the struggles witnesses go through in trying to find a conventional explanation for the unconventional event they've just witnessed, whether the sighting involves hairy hominids, UFOs, or both. In January of 2004, while my parents were with me at my house in Texas, my mother pointed out a bright light that was visible through the trees in front of the house. The time was about 9:00 at night, and the sun had long since set. As the branches waved in the breeze, the light popped in and out of sight. Its height suggested it lay beyond the houses across the road from my property, either on or slightly above the hill to the northwest. No lights had been visible in that area before. My father, a devout cynic, dismissed the light as nothing.

My mother and I decided to trudge outside, up the hill behind the house, for a better view. My father grudgingly accompanied us. Now we could see a series of bright lights in the distance, off to the northwest, but still they looked like nothing more than bright lights. My father suggested they were arena lights, bright bulbs on poles used to illuminate a riding arena after dark. Curious, my mom and I got into my car and drove down the winding roads toward the area where we had seen the lights. From the vicinity of the lights, just over the summit of a large hill, a glow brighter than the lights of a football stadium radiated up from a hollow in the woods. The light was visible for miles around, despite the hills and trees that masked whatever gave off the light.

We drove down a side road that paralleled the road on which I lived. We had just crested the hill when the source of the brilliance came into view—six huge lights aligned in two diagonal rows, three lights per row. Each light burned ten times brighter and larger than any football stadium's lights. I had the impression the lights hung from an enormous object, though the brilliance of the lights drowned out the object. The lights, and whatever they protruded from, hovered about 100 feet off the ground behind the main house and barn of a horse ranch. No lights burned within the house. From the barn, light seeped out from the rectangular hatches of the stalls. The ranch seemed eerily still, almost lifeless.

Before leaving the house, I'd felt acutely awake. Now, as I stared at the object in front of me, a lethargy infected my entire being. My limbs felt heavy, my mind numb. I kept thinking I shouldn't talk about what I was seeing. We drove a little further down the road, to a driveway barred by a locked gate. A sign informed us the land belonged to a gas company. Parked in the driveway, we continued to stare at the lights, which we could see through a break in the trees ahead of us. After a moment, we turned around and drove home.

At least I thought it was a moment.

We returned home in a daze, exhausted and lethargic. Soon, though, we realized a drive that should've taken 20 minutes had eaten up 45 minutes. I thought we'd sat in that driveway for a moment. Apparently, we had sat there in a stupor for 25 minutes. Neither of us could explain what happened to those 25 minutes. For a week I had an inexplicable urge not to discuss the incident, even with my mother.

People who encounter UFOs often comment on the psychological and physiological effects they felt—lethargy, inability to move—and sometimes thoughts will pop into their heads, ideas they know did not originate in their minds. Not everyone experiences the ideas as full-fledged thoughts. As in my case, often people simply feel the idea as a strong urge. Usually, those thoughts/feelings urge them to keep the UFO incident a secret.[65]

Remember the father of the Pennsylvania girls who saw a hairy hominid carrying a glowing orb? After he entered the woods to search for the creature the girls had seen, he returned in an odd state of distress—refusing to discuss what happened, telling others they shouldn't discuss it either. Perhaps he experienced the same psychological effects as UFO witnesses.

Most witnesses to hairy hominids experience psychological effects too. An acute sense of fear and danger overcomes them. Even people who know the wilderness and its denizens well, such as hunters, become overwhelmed by the sensation. Both the fear experienced by Bigfoot witnesses and the lethargy coupled with an irresistible urge to keep their sighting a secret felt by UFO witnesses may stem from the same phenomenon. UFO researchers have theorized that the beings who pilot the UFOs may employ microwaves to influence the minds of witnesses.

A handful of Bigfoot researchers have speculated the creatures possess the ability to use infrasound, very low frequency sound waves, to incite fear in witnesses. Both effects serve the same purpose—they discourage witnesses from talking. Both the fear and the inability to discuss the incident linger after the sighting itself has become history.

As for my UFO encounter, at first I accepted the arena lights explanation. But, as the days passed, I became less and less satisfied with it. A week later, my mother and I drove back to the area where we'd seen the huge lights. In the daytime, we searched the area for the lights' source, to no avail. The ranch had no arena, much less arena lights. The area behind the ranch had no poles of any kind, nothing atop which the lights could've sat. No roads or driveways or paths accessed the area. The sole residents of the locale were trees and bushes.

Since that night, I've sought alternate explanations—anything besides a UFO. I've come up with nothing. No one else I've spoken to has posited a theory that makes sense either. The incident remains a mystery. I've had other UFO sightings before and since that sighting. None has succumbed to a "rational" explanation.

Most of the other UFOs I saw in Texas consisted of bright lights in the sky. While driving down a country road near my house one night, I noticed a bright light off to the northwest, which at first I took for an airplane. As I watched, the light abruptly shrank to one-tenth its previous size. A few seconds later, the light winked out altogether.

I've watched airliners as they leave the international airport situated between Dallas and Fort Worth. When they reach a certain altitude, they shut off their bright lights, leaving on only the blinking lights. What I saw was not an airliner.

Mystery of the Black Cats

On December 28, 2003, at 11:30 PM, I was driving home down the same road on which I'd seen the bright light that shrank and vanished. As I passed a side road that dead-ended on my road, I spotted a dark shape just ahead of me on the left side of the road. The animal trotted around the corner onto the side road, disappearing into the night. The animal walked on four legs. Though I couldn't see the head or the tail, I recognized the gait as distinctly cat-like. The beast was too large to be a dog. Besides, I had become familiar with all the dogs that lived in the neighborhood, and none were black. Having spent my entire life around animals, particularly cats and dogs, I knew the animal I'd just seen moved with the fluidity and grace of a feline.

My father had also seen a black panther near my home. While driving to my house one evening, he glimpsed a dark shape up ahead, which he at first took for a dog. As the car's headlights illuminated the figure, he realized it was no dog. He had just seen a black panther.

Talking to other local residents, I discovered plenty of them had seen large cats too. Most had seen tan cougars, a.k.a. mountain lions, but some had seen black panthers. Conventional wisdom, as well as the Texas Parks & Wildlife Department, says Texas cougars died out in the 18th Century after hunters shot the last one. Perhaps a remnant population had survived, by some miracle, to resurface later. Black panthers, however, weren't supposed to be there at all. Perhaps some of the cougars had passed on a recessive gene that turned some of their offspring black (melanistic).

A bit of research exposed a dilemma. A few wildlife experts say melanistic cougars are rare but do exist. Most say cougars don't produce melanistic offspring. Jaguars do—except jaguars have been extinct in North America for hundreds of years. Mexico has some jaguars, and a few have migrated over the border into New Mexico and Arizona. But none should've made it to northern Texas.

I had also found the tracks of a big cat in the woods behind my house. They resembled not the cougar tracks I saw during my research into big cats but, instead, they looked like the jaguar tracks I'd seen. I had a real mystery on my hands.

And the mystery involved more than jaguars.

A Wolf in Wolf's Clothing

Over the years, my game camera has captured images of deer, rabbits, crows, and other sundry wildlife. I look forward to checking the camera each morning to see what if may have caught in its photographic trap overnight. On October 10, 2003, I had a surprise awaiting me on the camera's memory card.

As I thumbed through the photos, I saw the usual suspects—a.k.a. my dogs—as well as blank images in which whatever had triggered the sensor had scurried away before the camera snapped its image. Branches waving in the wind also seem to trigger the sensor. As I clicked through the blank images, however, a shape suddenly snagged my attention. At first, it looked like a dog. Then I realized it wasn't a dog.

A coyote? I couldn't tell by looking at the image on the one-inch-square LCD display on the camera. I took the camera into the house to view the images on my TV.

The canine in the photo was larger than a coyote. While I had unlocked the chain that held the game camera in place, one of my dogs had wandered

One October night this canine wandered in front of hte game camera; too large for a coyote, the animal most resembles a red wolf, a species supposedly extinct in the wild (Texas)—October 10, 2003.

in front of the camera, triggering the sensor. The dog—an Anatolian shepherd named Berry—stood in approximately the same position as the mystery canine. Anatolian shepherds are large dogs. Berry comes up to my knees.

I compared the two photos to estimate the size of the mystery beast. The canine who had visited during the night stood as tall as my Anatolian shepherd. The coyotes in my area were short and scrawny, nothing like Berry. Not only did the canine in the game camera photo stand as tall as Berry, he also looked as heavyset as the Anatolian shepherd. No gray wolves, a.k.a. timber wolves, live in Texas. As with the black panthers, conventional wisdom and the TPWD state that wolves died out in North Texas long ago.

Back to the research I went. My wolf didn't look like the gray wolves. He resembled red wolves. I traveled to a wildlife park southwest of Fort Worth to see their red wolves. Those canines looked identical to the animal in my game camera photo.

Big problem. Red wolves are an endangered species who went extinct in Texas a long, long time ago. Yet my photo seemed to show a red wolf.

Several times I heard wolves howling. On three more occasions—March 27, April 5 , and April 9—the game camera snapped images of wolves. Where did these wolves come from? How did they get to my home in Texas? How did the jaguars get there?

Perhaps the answer has to do with the other unexplained events at my home—the flying "thing" and the glowing orbs.

Step through the gates, past the bearded lady, into...

The Carnival

Someone had decided to transform the back woods of my property into a funhouse. How long the perpetrators had exploited my property in this manner, I can't say for sure. I never knew about the carnival of weirdness until I bought a game camera. Within two months after I bought the camera, strange things started to happen in front of its neutral eye.

Strange *things*—what else can I call them? The objects preserved on virtual film defy any and all explanation. They simply *are.*

Crystal Balls

October 14, 2003. Opening day for the carnival.

I picked up the game camera on my morning walk, and took it back to the house to examine the images. Flipping through the images on the tiny

LCD display had revealed nothing big, yet I always like to double-check by viewing the photos on my TV screen. You can miss a lot, from rabbits to the eyeshine of unknown animals, if you only look at the images on the built-in display.

To the house I went, camera in hand. As I scanned the images on the TV screen, I saw nothing unusual—until halfway through the night's collection. There, captured in mid-air at 7:23 PM, a glowing orb hovered in front of a tree. No animals frolicked in the foreground, or the background. Nothing populated the image except grass, trees, and the luminous sphere. Clear and in focus, eerie as a soothsayer's crystal ball, the orb radiated a bluish light from its mottled core, while its rim burned bright white.

I tried to imagine a scenario in which particles of dust could muster the impression of a glowing orb. But, if dust reflected the camera's flash, why did a single orb appear, instead of dozens? To think a single particle

The first orb captured by my game camera, shown in the original photo and in closeup at left (Texas)—October 14, 2003.

of dust could waft into the air to refract the camera's flash into a false orb strains logic to the brink of snapping. The previous night had been calm. What force could suck up dust without wind? The answer is...none.

The next day, I tromped out to the scene of the crime to gather any evidence that might identify the perpetrator, whether dust or something else. By standing at the tree onto which I'd strapped the game camera two nights earlier, I compared the photo of the orb to the tree directly across from me, the tree in front of which the orb had hovered. Then I took some measurements. The orb had appeared to hover right in front of the tree, just below a branch. Assuming appearances were correct, I estimated the orb measured four inches in diameter and hung 5′ 1″ off the ground.

On the night of November 9, at 8:31 PM, the camera shot a surreptitious image of another orb. This time, the orb hid amongst the drops of a light rain that fell that night. The orb, however, is quite distinct from the raindrops. The latter appear translucent, faint; the former appears solid, bright, and smudged with a slight motion blur. Two more possible orbs hover behind and to the left of the first (from the camera's viewpoint), though the companion orbs sit further from the camera, appearing fainter. Their identification as orbs

In this photo, a horse strolls by in the background as a bright orb hovers nearer the camera; the night was cold, insect-free, and calm (Texas)—November 20, 2003.

A deer seems not to notice the orb hovering nearby; these glowing balls often appear in photos with animals, as if attracted by them (Texas)—August 22, 2004.

rather than raindrops must remain tentative—unlike the first orb, whose brightness and motion blur rule out a raindrop as the culprit.

Since October of 2003, my game camera has netted countless pictures of orbs. In the most striking examples, the photos catch orbs in the same frame as an animal or animals. For instance, in a photo taken at 12:23 AM on November 20, 2003, a horse lies on her side before the camera, in the process of rolling (my horses like to lounge in front of the camera; the flash seems to attract them). Another horse pokes her head out from behind a tree to watch her buddy. Between the two horses, near the tree, an orb loiters. The orb appears in a single image. The previous and subsequent images illustrate the horse rolling in the leaves and dirt, without the orb monitoring. Another snapshot of a horse—taken at 10:54 PM on the 12th of December, three weeks after the incident just mentioned—included an orb too, this time much closer to the camera, brighter and more defined. The mottled center is clear, the rim brilliant.

Lest you should think only my horses attract these balls of light, on the night of August 22, 2004, the camera snapped a shot of a deer as it traipsed by the camera at 9:44 PM. Behind and to the left of the deer hovers a bright orb. My dogs have posed with the orbs too, on several occasions. The animals don't seem to notice the orbs, or else they have

The orb in this photo appears to have a motion blur (see the closeup at left), suggesting that a real, physical object raced across the field of view (Texas)—December 13, 2003.

become accustomed to them, so they no longer react to the presence of an orb or orbs.

The photos have demonstrated that these orbs both travel on their own power, regardless of wind speed/direction or lack thereof, and generate light internally. A series of photos, in which my dogs dawdle in front of the camera sniffing around, establish that an orb moved past them unnoticed. The orb first appears in the tree to the right of and beyond the dogs. In the next shot, the orb has zoomed closer, dipping toward the ground. In the last image, the orb is gone.

A photo taken in the early hours of December 13, 2003, shows a large orb whizzing through the frame with other, smaller orbs nearby. The large orb appears smudged at the lower-right corner. The smudge is motion blur. None of the other orbs in the photo display a motion blur. If the camera had gotten jostled, everything in the photo would look smudged. Yet only the single orb possesses a motion blur.

The Amazing Flying Paper Plate

On the morning of November 22, 2003, I retrieved the game camera as usual. Also as usual, I had installed the camera on the hill behind the house, in the woods along the northern trail. The previous night had been windless and warm, the kind of night that inspires animals to wander. I hoped a jaguar or wolf had moseyed past the camera for a close-up, or maybe a ball of light had floated past the lens. After browsing the snapshots on the camera's LCD display, I took the camera inside for a closer inspection of the images. On the TV screen, I surveyed the night's photos. The 13th image stopped me.

What was that thing in the lower right of frame? The white object hovered above the ground, casting a shadow on the dirt. It looked like a paper plate, battered and buckled along its rim. But, unless I've really missed out on some important news, paper plates don't fly. The object in the photo clearly flew, or floated, above the ground. Yet the previous night had been calm—no wind whatsoever, not even a breeze to tickle the dead grass. Nothing could have flown or floated past the camera unless someone threw it.

Or the object had an internal power source.

I showed the photo to a number of people, including an engineer. None could explain it. Someone suggested the object might be a plastic grocery-store bag. The same person had no answer when asked how a grocery-store bag could fly of its own volition. The previous image, taken in the same minute as the portrait of the flying thing, showed nothing in the vicinity.

The next photo in the night's sequence, taken four minutes later, caught my thoroughbred meandering down the trail. On the ground, about three feet from his right front hoof, hovers an orb of grayish light rimmed with white light. No insects stick around into November. The horse had not kicked up dust. He was walking too slowly to disturb the dirt enough for dust to go flying in all directions. Besides, if dust could reflect light in such a way that it created the illusion of glowing orbs, why would a lone orb appear in the photo? Why not several, or dozens?

Spook House

You'd think a flying paper plate would top the list of bizarre images, yet it falls short of the number one spot. That honor goes to the light show that took place on June 6, 2004.

June 2004 brought some of the fiercest thunderstorms in years. My sister's house had been hit by a tornado on June 1, though, fortunately, she and her children were away from home at the time. The storm that

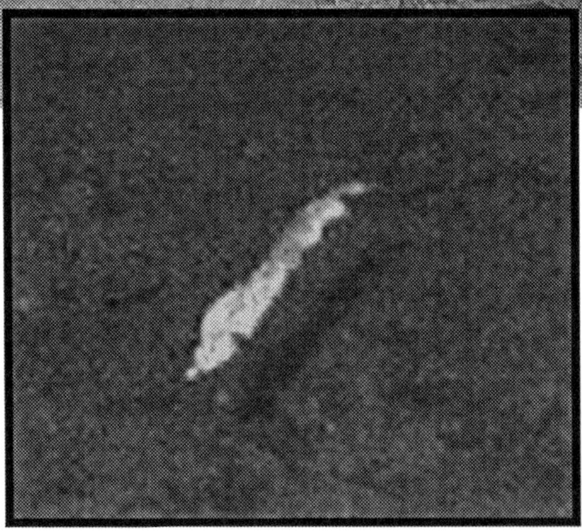

This object showed up in the middle of the night (Texas)—November 21, 2003.

The white object on the ground, at image center, is a salt block. Behind and to the left of the salt block, the eyeshine from a horse is visible. The mystery object is floating, or flying, above the ground, casting a shadow. The night was still, so no breeze propelled the object. As with many orbs, this object appears in conjunction with an animal.

struck her home, which lay fifteen miles from my property, besieged my home with large hail and hurricane-force winds so fierce the trees knelt down before their might—literally. Somehow the trees survived without snapping, though countless limbs shattered to the ground. The storms themselves behaved quite strangely. Meteorologists, both in my local area and on national networks like the Weather Channel, commented on the "tropical dimensions" of the storm. It spun like a hurricane over North Texas. The radar and satellite images showed the storm system spiraling so tightly it took on the suggestion of an eye, similar to a hurricane's center. Within the spiraling system, individual storms shifted their tracks. First moving from northwest to southeast, the storms later doubled back. I lived in Texas for fifteen years, yet I had never before witnessed a storm that behaved in such a way.

The events of June 1st rank as the worst, and most peculiar, storms I've experienced. The carnival in my back woods, however, had yet to peak. In fact, its producers had reserved their star attraction for the night of June 6th. What my game camera witnessed, and preserved on its memory card, that night surpassed anything the carnival troupe had staged before. That night another storm system ripped through the region. Ferocious lightning, wind, and torrential rains assaulted North Texas. I watched from my back porch as the game camera flashed in between the bursts of lightning. Still, I had no idea what the camera had seen.

The sensor detects motion and heat. Most people don't realize that vegetation gives off a small amount of heat. When the branches wave in the wind near the camera, they can trigger the sensor. I have dozens of photos in which the camera has captured the motion of branches. On the night of June 6th, however, the camera snapped images of more than branches flapping in the wind.

Before sunset, the camera took a cluster of images that featured nothing more interesting than trees, grass, and earth. I assume flapping branches triggered the sensor. But, at 9:51 PM, as sunset entombed the last relics of daylight beneath the horizon, the camera began clicking away. The first 13 images show darkness encroaching, rain splattering the camera's eyehole, the dampened earth surrounding the camera. Then—in the 14th image, shot at 9:54—a tiny orb appears at camera-left. Though the rain has fogged the camera lens, veiling the image in thin haze, the orb glimmers white against the dark backdrop of trees and earth. For the next two minutes, the camera takes empty images. At 9:56, an indistinct shape materializes low to the ground, some small animal whose eye refracts the light from the camera's flash. After several more pictures of nothing, at 9:57 a pair of orbs appear in the same area where the first had emerged. From that minute on, every image contains orbs, of varying sizes and intensities, in growing numbers.

The fainter, transparent objects are likely raindrops. But rain cannot account for the brilliant orbs that streak across the frames. And it certainly cannot account for the funhouse display that ensued.

By 10:00, orbs swarmed in front of the camera, flying to and fro like insects on speed. Perhaps I should reiterate here that the camera shot these images during a thunderstorm. Long before a storm hits, when the air stills and the landscape descends into a preternatural hush, every living thing heads for a hideaway—including the insects. While the rain would pummel the earth for a few minutes, then abate for a spell, the insects remained in hiding. The orbs are not insects. Nor are they dust particles, as dust cannot waft up from wet soil.

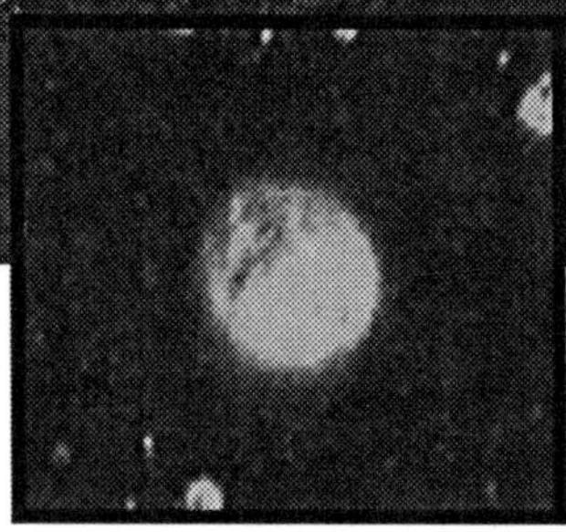

At 9:11 PM, this cluster of orbs showed up; many seem to be moving upward, and the one in the top center has a three-dimensional quality to its appearance (Texas)—June 6, 2004.

The image taken at 10:00 has some unusual characteristics. The orbs show a motion blur—going *up*, not down like raindrops. Unless a new kind of rain has evolved on Earth, the sort that rises from the ground into the clouds above, rain simply cannot explain this image. Raindrops streaking down the camera's window would blur *downward*. In the succeeding images, the orbs seem to be popping up out of the ground. The camera continued to click away, catching dozens, perhaps hundreds, of orbs in flight. Some of the biggest, brightest orbs possessed a three-dimensional quality—like freaky softballs hovering amid the rain. A picture taken at 9:11 illustrates this point. The orb in the upper center of the frame has the typical mottled center, bluish gray in color, with a brighter rim of white. Yet, unlike orbs in photos from previous nights, this orb has a singular three-dimensional appearance. Perhaps due to its mottling, the object resembles a bluish version of our moon!

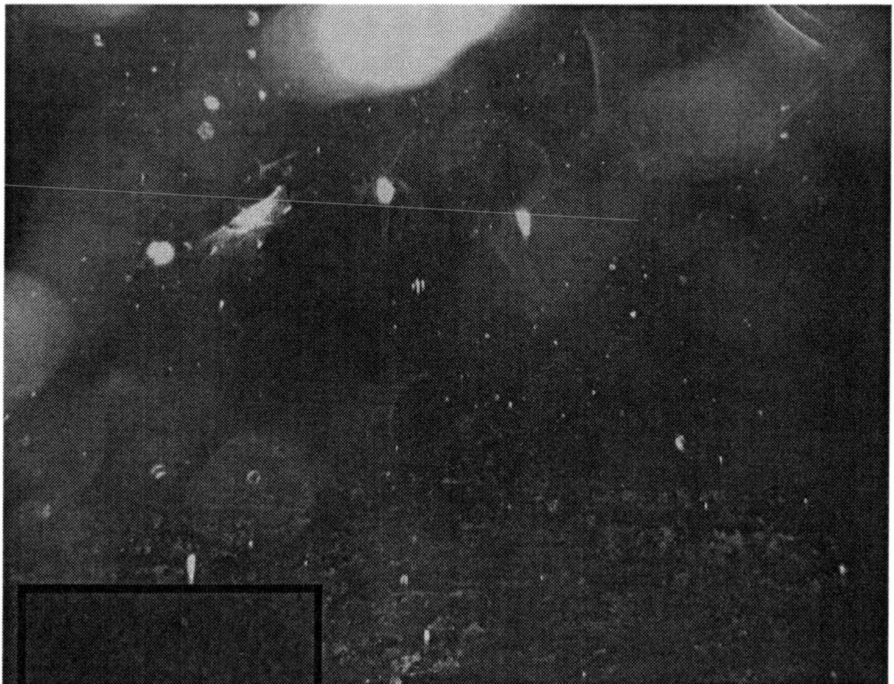

A strange configuration of lights on the game camera, the same night as the "moon" and other bizarre orbs (Texas)—June 6, 2004.

The show must be over, right? What could possibly outdo a tiny moon hovering in the woods behind my house?

Plenty. The spook house had yet to release its most freakish residents.

The light show had refused to subside. The game camera dutifully snapped image after image of orbs flitting past in eerie swarms. At 10:13, the swarm dispersed somewhat, making way for a new entrant in the freak show—the three-headed glowworm.

I can't think of any better moniker for the object (or objects). Near the ground, and perhaps on the ground, just off-center from the camera's viewpoint, a trio of rod-shaped lights huddle together. Their tops almost touch, so that they seem to form one object. The center rod bulges out at its top, something like a bowling pin flipped on its head. A similar trio of rod lights makes an appearance about a minute later, this time at the upper right of the image. In the second image, the rods have taken on a reddish tint and look as if they've drifted farther from the camera (if in fact they represent the same object). Meanwhile, the camera keeps filling its memory card with snapshots of orbs, orbs, and more orbs.

In this image from the night of the spook show, an orb appears to peek out from behind a tree while a third hovers in front of the trees (Texas)—June 6, 2004.

At 9:27, the camera catches an orb peeking out from behind a tree (or ducking behind it, take your pick), on the right-hand side of the image. The orb is partially hidden by the tree. The camera stood at least 20 feet from the tree, which makes the orb very large.

If I had any doubts left that the orbs can generate their own light, the photo taken at 11:52 that night eliminated it. The camera did not use its flash, yet it took an image of glowing orbs. In previous images, when the flash did go off, the ground is visible. In this image, everything is dark—except the orbs. At 12:01, the camera once again fails to use its flash. Yet, there again, the orbs cavort before its lens.

By 1:00 AM, the show had ended. While the camera took a few more pictures, nothing else glimmered before its lens. By 1:39, the camera had given up too. It recorded no more images that night.

Witness to the Revelry

The events of June 6, 2004, proved unique in their intensity. But the phenomenon itself, the glowing orbs, has never stopped completely. Since I moved north, my game camera has captured more orbs, including one instance where an orb hovered near a deer.

Just as the orb mystery has endured beyond my cross-country move, my UFO experiences have persisted too. On February 28, 2005, at 9:00 PM, I wandered over to the window above the landing on the second floor. From there, I have a good view of the skies above the treetops. As I observed the stars, I noticed a bright light just above the trees, between the crowns of two pines. The light measured about five times larger than even Jupiter. Its color was a dull orange. I stared at the light for 10-15 seconds, then glanced away at the stars elsewhere in the heavens. When I looked back a second later, the light had vanished. This area has no major airport. The Houghton County Airport is small, with a single airline offering commuter service into and out of Minneapolis. The large airliners that pass overhead remain high in the sky—you can't even hear their engines. When I spotted the orange light, I heard no sounds. It never moved. The light simply...vanished.

A handful of photos from the game camera caught orange balls in flight. Around the world, the orb phenomenon is known to include both bluish-white and orange balls.

I'd like to conclude this section with another eyewitness account, a close-up sighting that happened a few months ago.

On the night of June 5, 2005, I was awakened by a flash outside the bedroom window. The light flashed so bright it penetrated the veil of my eyelids. I glanced out the windows at the front yard, where the flash had seemed to originate. A line of thirty-foot cedar trees, standing about 20 feet from the front deck, screens the house from the driveway. In front of the trees, an orb pulsed. The light was bluish-white, with a brighter white rim. The orb glowed for a few seconds, then extinguished. A moment later, another orb (or perhaps the same one, which had moved) pulsed on for several seconds. To the right of where the orbs pulsed, I saw a swarm of fireflies flickering. I estimated the orbs to be similar in size to a basketball, far larger than the fireflies flitting about nearby.

The orb I witnessed looked identical to the orbs captured in the game camera photos.

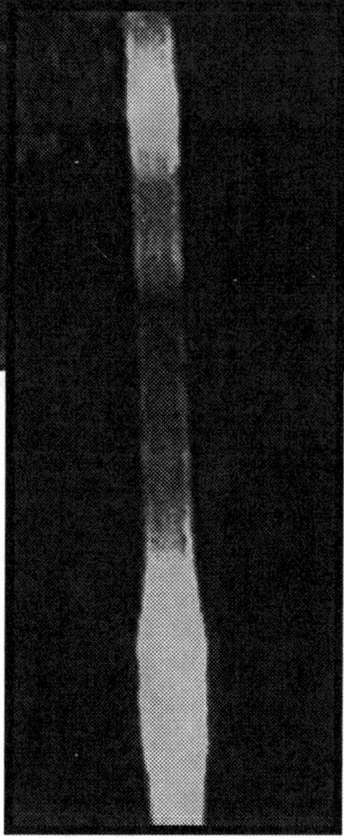

Late at night the game camera captured this bizarre pillar of light, the first of three times the camera would catch such an object; in the close-up at left, notice the balls of light visible inside the pillar, above the object's bright nucleus at the bottom of the image(Texas)—May 15, 2004.

Do the pillars bear a relationship to the orbs? The balls within the pillar do resemble the orbs photographed on other nights. Perhaps the orbs emerge from the pillars. I can't say for certain, and neither can anyone else.

In the top photo, an orb appears in the background near a tree. In the bottom photo, taken seconds later, the orb has swept closer to the camera. The dogs seem unaware of the orb, or have become so accustomed to seeing orbs that they no longer pay attention to them—although one dog has lifted her head, perhaps attracted by the flash of the orb, or some sound the orb makes that only an animal can hear.

(Texas)—November 10, 2003

On the night of the spook show, the game camera took several photos without using its flash, as evidenced by the fact that you can't see the ground or trees; this photo was taken without the flash, yet orbs glow in the frame anyway (Texas)—June 6, 2004.

A bright ball of light; I had put the game camera in front of the house, across the driveway from the porch, which places this light in the front yard or perhaps in the trees north of the yard (Texas)—July 17, 2004.

6
Tying the Knots

The threads dangle before us—from our DNA, from the fossils, from the tapestry of hairy hominid sightings and ancient mythology. Each thread represents a piece of the garment. To reconstruct the truth, we must knot the threads together.

Bigfoot Meets ET

Encounters with Bigfoot go back deep into history. Sightings of UFOs extend equally as deep into humanity's past. As with the ancient artwork, however, the ancient sightings cannot be corroborated or investigated. The witnesses have long since died. The scenes of the events have changed. Still, the stories the witnesses told offer a glimpse into the past, and possible clues to the future, that deserve our attention.

Ancient people were no less intelligent or erudite than modern people. Ancient monuments like Stonehenge and the pyramids of Egypt evince the ancients' understanding of the heavens. Artifacts also confirm for us that ancient people possessed a keen interest in, and grasp of, the sciences—from astronomy to metallurgy to engineering. As early as 15,000 BC, humans may have noticed the constellation we call the Pleiades. A cave painting at Lascaux, France, depicts six points that may represent the six visible stars of the Pleiades (the seventh star is difficult to see with the naked eye). The Nebra Sky Disk, found in Germany during the 1990s, illustrates the moon at its full and crescent stages, alongside a sky-full of stars—and the seven Pleiades. The Nebra Sky Disk dates to around 1600 BC.[57]

As mentioned earlier, myths concerning the Pleiades abound worldwide. Besides the commonality of the "seven sisters" association, these myths also feature a fascinating correlation with hairy hominids. In all the stories, a hideous giant chases seven sisters into the heavens, where they take refuge from the ogre by transforming into stars. The Greeks, who called the giant Orion—the namesake for the constellation—said that he continues to pursue the daughters of Zeus from his heavenly station. The aborigines of Australia tell a similar story, in which the ogre is represented by a star in Orion's belt. The Kiowa Indians of North America tell of a giant bear stalking the sisters. To help the women, the gods erected a stone tower (Devil's Tower, Wyoming) on which the women could hide. But the bear persisted, forcing the women to flee into the heavens.[57]

We've examined the legends of giants and other hairy creatures from antiquity. How interesting that an astronomical legend should include similar creatures. Myths of star gods flourished throughout antiquity too, as seen in the Pleiades stories. Doubly interesting, then, that the Pleiades myths speak of giants alongside the star gods.

Clearly, ancient people were familiar with celestial bodies. They knew the natural world as well. The exquisitely-rendered bison, bulls, horses, mammoths, owls, and other creatures from European cave art demonstrate the ancients' familiarity with the fauna of their time. In locales around the world, ancient people bequeathed to us artwork that discloses their knowledge of the natural world. They lived by nature's whims. Survival required an intimate knowledge of weather, animals, and plant life. So, when these same people tell us of encounters with strange craft and hairy bipeds, why shouldn't we listen?

After all, many myths have a basis in reality. For centuries everyone assumed the legend of Troy, as related by Homer in the *Iliad*, was pure myth. In the 19th Century, when archaeologist Heinrich Schliemann unearthed the real Troy on a hillside in Turkey, the myth became reality. Perhaps the truth upon which the ancients based their hairy hominid-UFO stories still awaits discovery.

Before I tie the threads together, I must explore one more lead.

Gods of the Stars

Around the world, humans have used myth to explain what they cannot account for in everyday language. Hence, we have the Greek myth of Demeter and Persephone to explain the seasons. Demeter's daughter, having eaten the forbidden fruit of the pomegranate while in Hades' kingdom, must spend half the year aboveground with her mother, the harvest goddess, and the remainder in the underworld. Demeter's anguish over her daughter's annual departure triggers autumn, when the living world descends into bleakness. When Persephone returns, spring blooms.

Few would argue that the myth of Demeter and Persephone helped the ancient Greeks understand the seasons. But, when we turn to the ancient star-god myths, quarrelling ensues. Why? Perhaps because to think the ancients had contact with ufonauts, who may have influenced or even directed our cultural growth, is simply too frightening to contemplate. Yet the myths take us back deep into antiquity, to a time when most written records took the form of religious treatises and artwork existed mainly to glorify gods or god-kings. Only by couching their stories in religious metaphors could the ancient writers convey information.

UFOs have coexisted with humans as far back as the arts allow us to trace our history. Cave paintings in Europe depict aerial objects in a plethora of shapes, including the stereotypical "flying saucer." Aboriginal art in Australia, much of which dates to tens of thousands of years ago, shows similar crafts, as well as beings clad in getups that are indistinguishable from modern spacesuits. In the southwest US, ancient peoples bequeathed to us images of elongated figures with bug-like eyes, known to archaeologists as the Barrier Canyon style of anthropomorphs. The figures bear a striking resemblance to modern space-suited astronauts. In a previous chapter, I suggested these figure represent hairy hominids. If hairy hominids share a connection with UFOs, the Barrier Canyon figures could represent both space-suited astronauts and hairy hominids—the creatures themselves could have been astronauts at one time, working for the star gods.

A discussion of space-suited beings and flying saucers depicted in ancient artwork could fill hundreds of pages. The brief examples presented above offer an overview. They also point out an interesting correlation—the same cultures that represented hairy hominids in their artwork *also represented UFOs in their artwork.*

What should we make of this? Perhaps the myths of the star gods can guide us, for such myths extend as deeply into humanity's past as legends of hairy hominids.

Sumerian writings talk about gods who came from the heavens. Ancient Egypt had star gods too. The god Osiris, ruler of the kingdom of the dead, was associated with the star Sirius. Some of Egypt's oldest religious writings, a collection of "utterances" known as the Pyramid Texts, compare the king to a star and describe how he will ascend to the heavens at the end of his reign. In Utterance 248, the text says the king is a star that travels great distances to "bring products" to the sun god Ra.[27] Egyptologists, and even most alternative scholars, interpret the Pyramid Texts as esoteric religious allegories.

Why should we not take them at face value? The ancient Egyptians produced some of the most stunningly realistic sculptures of humans and animals, equal in quality to Classical Greek statuary. The Egyptians employed rock crystals as eyes in many statues, painting on the backs of the crystals to mimic the luminosity and depth of human and animal eyes. If they went to such lengths to create realistic artwork, why should we presume their writings are incomprehensible metaphors?

Of course, the major problem is that all ancient cultures have become exotic and mysterious to us. With only writings and artwork by which to judge them, we can easily fall into the trap of ascribing mystical wisdom to everything they did. If we examine the ancient evidence with clear eyes and a mind willing to accept the evidence without over-interpreting it, what do we see?

The Pyramid Texts go on for 759 "utterances" (sections consisting of one or more paragraphs). Utterance 261 narrates how the king travels into the sky, navigating the globe as a "flame"—usually interpreted to mean a bolt of lightning. This "brilliant" king who moves in the sky, high above the ground, could literally refer to a bright light inside of which the king embarks on an aerial journey. Hmm...a bright light, like a flame or star, in which the king travels...sounds kind of like a UFO!

In Utterance 302, the king "flies" into the heavens like a bird to join the gods. Before brushing this off as a metaphor for his soul going to heaven, pause for a moment to think about the wooden sculpture found in an ancient Egyptian tomb. The sculpture looks like a model airplane. Tests have shown that the model flies too—the craftsman who fashioned it made the "sculpture," which Egyptologists prefer to think of as a stylized bird, aerodynamically sound.[15]

Countless cultures around the world have worshiped star gods, from the Maya of Central America to the Sumerians of the Middle East. The same cultures depict UFOs in their artwork, venerate religious texts that recount journeys into space, and preserve tales of hairy hominids.

If we assume myths reflect reality, we're left with a puzzling question. What connects UFOs and hairy hominids?

Teasing the Puzzle

No one can answer the question without engaging in speculation. No one can claim to have found the final, unequivocal answer. However, one can examine the evidence, align the pieces, and attempt to tease a coherent picture from the puzzle. This is what I've endeavored to do. Now, I will outline what I see as a good candidate for the answer.

First, let me reiterate the salient points made thus far:

- ✧ Hairy hominids are not the only out-of-place creatures found in the woods; red wolves and jaguars have reappeared in the same areas inhabited by hairy hominids.
- ✧ UFOs frequently appear in the same areas where these creatures live.
- ✧ Witnesses have seen UFOs in conjunction with hairy hominids.
- ✧ The worldwide star-god myth of the Pleiades mentions giants who chase seven sisters into the heavens.
- ✧ The *ikals* of Mexico are said to fly with the aid of rocket packs, while the *farfadets* of France have "flying chariots."
- ✧ Cultures around the world, and throughout history, have worshiped star gods and written down myths of flying objects and people who ventured into the skies with the gods. Those same cultures have legends of hairy hominids.
- ✧ Those cultures also preserve tales of star gods influencing or kick-starting human life on the planet.
- ✧ All humans on Earth are more similar genetically than two gorillas living in the same forest.

Wildlife biologists say that red wolves and jaguars went extinct in Texas long ago, yet today they thrive in the Lone Star State. Evolutionary theory requires that the ancient hominids disappeared soon after modern humans emerged, yet their descendants live on in the guise of Bigfoot and other hairy hominids. Remember that when I say "descendants" I mean *genealogical* descendants, not *evolutionary* descendants.

No evidence exists of an intermediary stage between apes and hairy hominids. One moment we have primates, the next hairy hominids. While paleoanthropologists insist upon dividing the ancient hominids up into countless species, the differences between the three-foot *Australopithecus* and the six-foot *Homo erectus* could just as easily represent diversity within a species. Modern humans display a similar, if not greater, amount of physical diversity. Fossils cannot tell us where to draw the line between species. Even today, science has no definitive answer to the question of what makes a species.

With humans, the puzzle becomes more confusing. Human beings sprang up out of nowhere, millions of years ago. Our DNA shows that we are more similar genetically than the members of any other species. Perhaps the myths of star gods creating us do have some basis in historical fact. We've lost the historical records that tell us how, why, and when that act of creation happened. We hardly need a vivid imagination to see how such records can vanish from history.

Lost in Time

How many times in our *recent* past have we humans obliterated our own history? The perpetrators of the Spanish Inquisition murdered people who believed in the pagan gods, and destroyed every written record they found of the pagan beliefs. Thus, while wicca (the religion of witchcraft) has existed for thousands of years, the version of it practiced today may bear no resemblance at all to the ancient version. The practices and beliefs of the ancient wiccans were erased.

The Spanish accomplished the same mission in Central America. The Conquistadors destroyed volumes of codices, the written records of the Mayan culture, leaving only a handful untouched. The burning of the library at Alexandria, in Egypt, destroyed an untold wealth of knowledge culled from the ancient Egyptians, Greeks, and Romans. The Sumerians, Babylonians, Romans, and other cultures from the ancient Mediterranean boasted similar libraries which were eradicated by war or time. Who knows what else we have lost.

The evidence for high culture before the "dawn of civilization"—around 5000 BC, by conventional wisdom—has piled up to quite a height in recent years. The December 2004 tsunami that ravaged Indonesia also peeled back the layers of earth that veiled the "lost city" of Mahabalipuram on the Indian coast. The ruins include vast temples constructed from rectangular blocks of stone.[37] Other "lost cities" may still languish beneath the waves, inundated when the sea levels rose catastrophically at the end of the last ice age. Divers have glimpsed underwater ruins off the coasts of Japan, Malta, and Egypt. Archaeologists once dismissed the tales of a sunken city off India's coast as hogwash, exactly as they still dismiss the possibility of similar ruins offshore of other nations.

The last ice age ended around 10,000 years ago. By 9,000 years ago the first cities emerged. Catalhoyuk, in Turkey, had swelled into the largest Neolithic ("new stone age") village known in the archaeological record, housing 7,000-10,000 people. During the ice age, humans are presumed to have lived in diffuse tribes, wandering the ice-free zones without cohering into societies. Yet, less than a thousand years after the glaciers retreated, a Neolithic metropolis had built up on the plains of Turkey—a region swathed by ice until the big meltdown.

The residents of Catalhoyuk erected mudbrick houses packed together mere inches apart, packed so densely the people had to climb in through the roofs. The houses featured reed roofing, ovens, hearths, and platforms or walls to segregate the interior into distinct spaces. The inhabitants often decorated the walls with vibrant artwork, as well as leaving behind clay figurines, carved bones, and obsidian mirrors.[26]

The evidence is clear. Humans have had culture further back in time than archaeologists care to admit. How could people develop urban centers so quickly, relocating from their old homelands to the newly exposed plains of Turkey? How did they know where to go?

Human cultures have a tendency to spring up out of nowhere, then vanish or regress with equal abruptness. The Anasazi, the ancient Maltese, the residents of Teotihuacán in Mexico and Tiahuanaco in Bolivia—all these cultures sprang up, flourished, and vanished without a trace. They also built enormous, megalithic structures whose purpose and construction techniques remain a mystery. Many of these cultures preserve legends of strangers who came from across the sea, bringing with them civilization and technology the likes of which locals had never seen.

In South America, the civilizers were called the Viracochas, named after their leader, Viracocha. In Central America, Viracocha was known as Quetzalcoatl. The "temples" of Tiahuanaco include pieces of iron set inside the joints to strengthen the bond. The stones, nearly all of megalithic stature, are fitted so precisely a sheet of paper won't fit between them. A sheet of gold-plated iron was also found deep inside the Great Pyramid of Egypt, between the joints of its megalithic blocks.[34] Gold-plated iron *4,500 years ago?*

Few people realize the extent of the technology the Greeks and Romans possessed. The Romans had developed waterproof concrete that would set underwater. They built thousands of miles of aqueducts that could carry water over mountains. Their bathhouses were heated by hot water running under the floors and in the walls. They had toilets too, as well as advanced surgical techniques that included cataract surgery. Greek inventors developed the first coin-operated vending machines, automatic doors, steam engines, and countless other inventions. All these innovations died along with the civilizations that created them.

Garbage

The human genome contains enormous amounts of "junk" DNA—genes which seem to serve no purpose. How did we get all that junk inside us? Why should evolution preserve useless genes? Evolution by natural selection requires that beneficial mutations be perpetuated. What benefit comes from junk DNA?

We humans have a number of anomalies in our systems. In nearly all animals except humans, the brains of offspring reach their full size before birth. Yet in humans, babies cannot grow their brains fully before birth, because they would never fit through the mother's pelvis if they did.[63] This means our babies emerge as helpless beings requiring the constant nurturing and supervision of the parents. No apes have this scenario. If we evolved from apes, why did evolution give us this decidedly unbeneficial method of childbirth?

What about our hairless bodies? Having no hair forces us to wear clothing, build shelters, and warm ourselves by fires. In extreme heat, we have no defense. I've just experienced the hottest summer in years here in northernmost Michigan, a heat we endured without air conditioning. I can tell you that even shedding your clothes does nothing to alleviate the heat! You sweat, which drives you to drink huge amounts of water. For ancient people, who supposedly had no plumbing or wells, this means they would've had to seek out water constantly in order to quench their never-ending thirst during the summertime. In Africa, even 100,000 years ago, summer was hot. Yet the best reason paleoanthropologists can give for our lack of hair is this: we lost our hairy coats so that we could shed heat via our skin. Why then do northern peoples, like Eskimos and Scandinavians, not have hairier bodies than Africans? If hairless bodies evolved as response to heat, those of us living in northern climes should have hairier bodies. Yet we don't.

We humans also sunburn. Again, if hairless bodies benefited us, we should have no trouble with the sun. Our skin does protect us somewhat by tanning, but eventually we will burn. What benefit does sunburn have for us? If evolution sold us these bodies, I want a refund.

Humans have bodies ill-adapted to the planet we inhabit. Our advanced societies rise and fall throughout history. Technology is attained, then lost. UFOs are depicted in the earliest artwork, alongside hairy

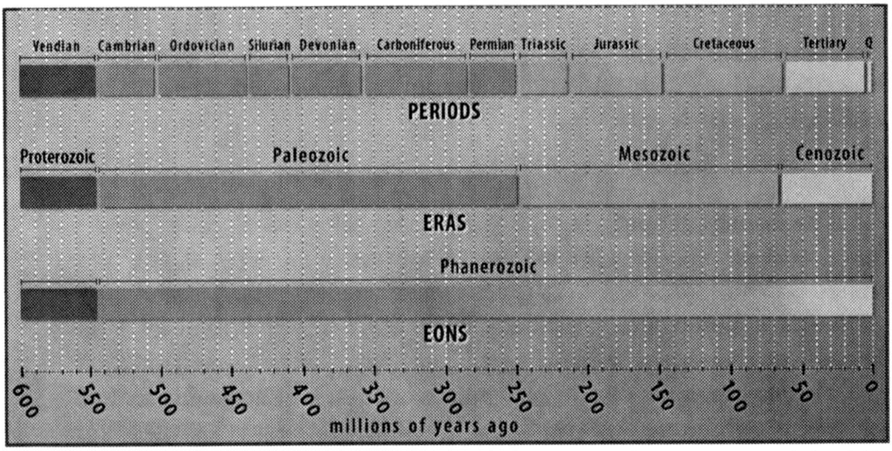

A timeline of the geologic eras and periods, based on the US Geological Survey's geologic time scale (available at geology.er.usgs.gov/paleo/geotime.shtml); "Q" represents the Quaternary Period, the current epoch.

hominids. Myths of civilizers from distant lands show up in cultures across the world. What's going on here?

Reverse Ark

I've talked about the jaguars and red wolves, animals that went extinct yet still wander the woodlands of Texas. Similar mysteries crop up when you examine the evidence from prehistory, a time from which no written records survive. Let's hop in our time machine and travel way, way back to witness the cycle of birth and death that serves as our planet's heartbeat. I will use the geologic time scale endorsed by the US Geological Survey, available online at http://geology.er.usgs.gov/paleo/geotime.shtml.

About 90% of Earth's history falls into Precambrian Time[29], a series of four eras that account for the time between the birth of our planet 4.5 billion years ago and the onset of the Cambrian Period 544 million years ago. The first life, according to orthodox science, took the form of single-celled organisms. These organisms grew into multi-celled organisms that in turn evolved into ever more complex life forms as the ages crept past. Complex life first appears in the fossil record during the Cambrian Period, which encompasses the period from 544 to 505 million years ago. From this point, evolution tells us, life should've slowly progressed into ever more complex organisms. For awhile, life does seem to have followed this program—until everything exploded.

The geologic moment at which this happened is often referred to as the Cambrian Explosion, because it occurred during the Cambrian Period. In an evolutionary second, life burgeoned into abundant and varied forms, creatures so strange we struggle to understand them from their fossils, which mainly consist of impressions left in the rock. The Burgess Shale formation in Canada contains the fossils of marine creatures from the Cambrian, whom scientists theorize perished in an undersea landslide. From trilobites to sci-fi movie creatures, the Cambrian seas teemed with life never seen since. A catastrophe, perhaps a meteor impact or volcanic cataclysm, wiped out the Cambrian creatures about 440 million years ago, at the end of the Ordovician Period. Another, lesser extinction followed at the close of the Devonian Period, about 360 million years ago.

At 248 million years ago, the "mother of all extinctions" wiped out 95% of all marine life on Earth and devastated land animals as well. This event, the Permian extinction, marks the boundary between two eras—the Paleozoic and the Mesozoic—and ushers in the Triassic Period. Miraculously, life explodes again in the wake of the Permian cataclysm. Fate intervenes once more, with the Triassic extinction around 213 million years ago, a vastly smaller event than the Permian extinction. The dinosaurs make their appearance during the Mesozoic, ruling the planet until another cataclysm eradicates them at the conclusion of the Cretaceous Period about 65 million years ago. Scientists refer to this extinction as the K-T boundary (Cretaceous-Tertiary; Cambrian gets the "C" designation), for a visible border separates the two periods in geological strata around the world.

We've taken a whirlwind tour of life on Earth. The twofold point to grasp here is the cycle of mass extinctions coupled with the odd manner in which life survived and, indeed, flourished in the aftermath. Debate still roils over what caused any of the mass extinctions. No one can explain how life rebounded after each either. Mass extinctions have given us another enigma too—the riddle of the "Lazarus" species. These creatures disappear after a mass extinction, sometimes remain absent for millions of years, and suddenly rematerialize. After the Permian extinction, corals vanished. Tens of millions of years later, during the Middle Triassic, corals reemerge as abruptly as they had disappeared.[19]

What happened? The evidence could be explained if we assume an outside force has manipulated life on our planet from the outset. In fact, we can find clues to this in the Christian Bible. Remember the story of Jacob and Esau?

The Bible describes Jacob as a normal, smooth-skinned man; his brother Esau is covered in a coat of reddish hair. While pregnant with the two boys, Rebekah receives some strange information from God—He tells her she has "two nations" growing inside her, two different kinds of men. If

Jon Bodsworth / The Egypt Archive

Scultpure of an Amarna princess, probably a daughter of the pharaoh Akhenaten, revealing the elongated cranium (Egypt)—14th Century BC

Jon Bodsworth / The Egypt Archive

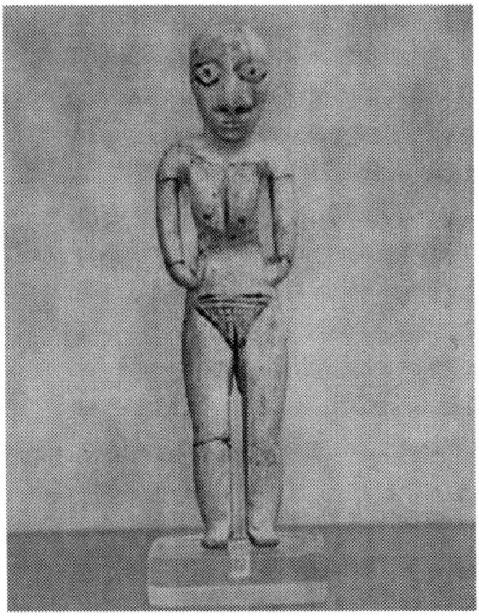

A Predynastic sculpture, exhibiting some strange, alien features (Egypt)—5th to 6th millennium BC

we interpret the text literally, then God has told Rebekah she will give birth to two children, one of whom is not quite human. Only two options exist to explain how a human woman could give birth to a nonhuman child: either she interbred with a closely related species, or she was artificially inseminated.

One modern story tells of a female hairy hominid being captured in Russia during the 19th Century. Though she could not speak, she lived among the locals and, supposedly, gave birth to half-human children whose descendants lived on into the 1950s. The descendants of the hairy hominid, named Zana, possess hairier bodies than most people, as well as darker skin than the average Russian. One grandson of Zana could reportedly lift a chair, complete with a man sitting on it, by grasping the chair in his jaws.[33]

The Bible tells of Rebekah marrying a human man, Isaac. Besides, she gave birth to *two* sons, only one of whom was hairy. The boys must've had different fathers, which rules out natural means. Only artificial means could produce two children with different fathers.

I have to bring up one last issue before outlining my theory. Around the world, from the Nazca plains of Peru to Kow Swamp in Australia and even ancient Egypt, skulls have turned up whose morphology both mystifies and disturbs us. The deformed skulls take several shapes: elongated types that seem to have gotten stuck in a vacuum hose; flattened skulls that look as if you placed your hands on either side and pushed really hard; and various kinds of lumps and protrusions that misshape the bone. Most often, several of these anomalies occur in one skull.

I've already mentioned the Kow Swamp skulls, which have flattened foreheads and elongated crania. Researcher Lloyd Pye has undertaken a project to analyze and understand a skull that came into his possession in the late 1990s—a bulging, bilobed skull with misshapen eye sockets that he has dubbed the Starchild Skull.[51] In 1907 in Egypt, inside a crypt designated Tomb 55, archaeologists unearthed the bones of a mummy whose skull was elongated.[52] The bones date to the time of Akhenaten, a strange pharaoh who outlawed the traditional gods of Egypt in favor of a single sun god. Artwork from Akhenaten's time depicts members of the royal family as having elongated heads, a portrayal archaeologists have tried to dismiss as artistic license, in spite of the Tomb 55 skull.

The same cultures that bequeathed to us these misshapen skulls have also left us strange, alien-looking sculptures. Take, for example, the sculptures from Egypt's Predynastic Period, a time before pharaohs took control of the dual kingdoms known as Upper and Lower Egypt. The Predynastic sculptures have oddly shaped torsos, heads slightly too large for their bodies, bug eyes, and misshapen hands. The figures most resemble the so-called "grays" described by modern UFO abductees.

Jon Bodsworth / The Egypt Archive

A bust of Tutankhamen, son and successor of Akhenaten; note the elongated skull, similar to the Paracas skull below (Egypt)—14th Century BC.

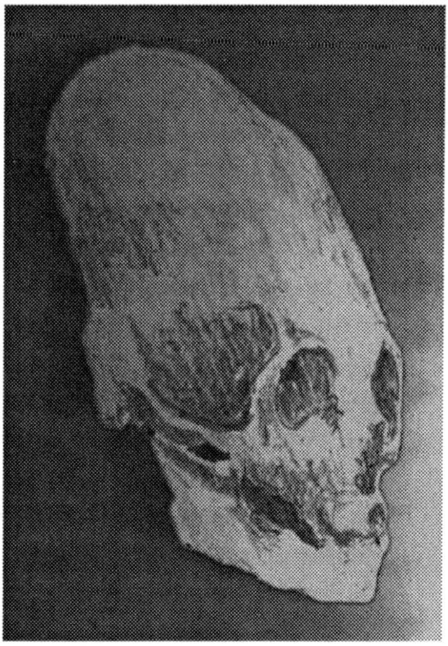

Archaeologists say this elongated skull, and others like it around the world, was deformed by skull binding (Paracas, Peru).

Time to recap the clues:

- ✧ Life explodes in an almost literal manner during the Cambrian Period.
- ✧ A cycle of mass extinctions beats life back, yet life regains its foothold after each cataclysm.
- ✧ Lazarus species vanish for millions of years, only to pop back up without warning.
- ✧ Ancient hominids burst out of the ape mold 4-5 million years ago with no precursors and no evidence of gradual evolution.
- ✧ Extinct species have reemerged today without explanation, like the red wolves and jaguars in Texas.
- ✧ Bizarrely shaped skulls appear around the world.

The clues lead me to one conclusion: someone has instituted a program of genetic manipulation on Earth, a sort of reverse ark in which—rather than taking species in to preserve them, as Noah did—species are created and set loose on the planet. Let them go, see what happens. Oops! A big asteroid wiped them out. Time to start all over, maybe try something different.

We take on similar projects today. Red wolves have been bred in captivity. In Australia, scientists are endeavoring to clone the extinct thylocine, a.k.a. the Tasmanian tiger, to repopulate the species. Perhaps Michael Crichton had it right in *Jurassic Park*. We do want to play God.

But someone else beat us to it.

What does a reverse ark have to do with lost technology or the ebbs and tides of life on Earth? How do star gods fit into the theory? Any civilization capable of creating life on such a monumental scale must possess technology so far beyond what we have today that it would seem like magic, like the work of gods, like supernatural forces at play. Perhaps some of these "beings," for wont of a better term, thought we needed a boost, technologically speaking. They gave us bits and pieces of technology to help us survive and grow. The global floods obliterated the ice-age civilizations, while later catastrophes wiped out other cultures. Along the way, kinsmen of the "civilizers" might've grown angry with their brethren for handing out technology to barbarians. Perhaps they destroyed the surviving cultures, by one means or another. One need not start a war to destroy a civilization. More insidious methods can work equally well.

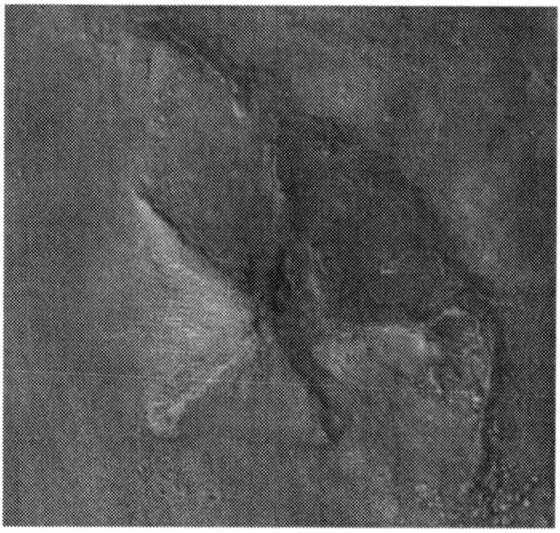

NASA/JPL/Malin Space Sciecne Systems

Is this a pyramid on Mars? While a definitive answer remains elusive, public interest in the Cydonia region endures. This image was snapped by NASA's Mars Global Surveyor in 2003.

Whether the "star gods" who kick-started life—and have controlled its progress ever since—represent aliens, residents of another dimension, time travelers, or something else completely, I can't say. No one can. Until our controllers announce their presence to the world, we can only guess as to their nature and intentions. It seems apparent, though, that their intentions are not benevolent—at least not entirely.

Perhaps clues do exist that can point us in the right direction, if only we knew how to interpret them. Consider the so-called "monuments" on Mars—the face, the pyramid-like structures, the "city square"—which all reside in the Cydonia region of the Red Planet. Various NASA probes have photographed the region, providing contradictory evidence as to the reality of the monuments. While the "face" seems to have been nothing more than shadows, the pyramid-like structures could prove more interesting. The enigma remains unsolved, leaving the tantalizing possibility that an ancient and long-dead civilization did build ruins on Mars.

I can't prove my theory. No one else can prove theirs either. Until time travel becomes a reality, nobody can know with 100% certainty what happened in the remotest past. Paleoanthropologists use bits of bone and heaps of conjecture to formulate their ideas about our identity as a species and from whence we came. I've chosen to look at the evidence gleaned from our ancient ancestors themselves. What they told us lies in the annals of history for all to peruse. How we treat it, whether we discount it or paint with the brush of religious symbolism, determines what we learn from the evidence.

The star gods have not left. What they have planned for us now, only they know.

References

1. Ashliman, D.L. "The Norse Creation Myth." D.L. Ashliman's FolkTexts, 7 Jan. 2003. <http://www.pitt.edu/~dash/creation.html>
2. Bahn, Paul G. & Jean Vertut. *Journey Through the Ice Age*. London: Seven Dials, 1999.
3. "Bigfoot Allegedly Spotted In Canada's Yukon Territory." NBC4 online, 15 Jul. 2005. <http://www.nbc4.com/news/4727069/detail.html>
4. "'Bigfoot' Hair Turns Out To Be Bison." NewsWatch 50 online, 28 Jul. 2005. <http://www.newswatch50.com/news/state/story.aspx?content_id=1B32D3D5-6D7C-45BC-B741-A1111E7F916B>
5. Bigfoot Roundtable. Coast to Coast AM with George Noory, broadcast 3 May 2005.
6. Boirayon, Marius. "The Giants of the Solomon Islands." Solomon Giants website, n.d. <http://www.thesolomongiants.com/>
7. Bord, Janet & Colin Bord. *The Bigfoot Casebook*. Harrisburg, PA: Stackpole Books, 1982.
8. Bord, Janet. *Fairies: Real Encounters with the Little People*. New York: Dell Publishing, 1997.
9. Bower, Bruce. "Lucy On The Ground With Knuckles." *Science News*, 157 (2000): 235.
10. Buckley, Archie. "Report on Sasquatch Field Findings." *The Sasquatch and Other Unknown Hominoids*. Ed. Vladimir Markotic & Grover Krantz. Calgary: Western Publishers, 1984. 187-201.

11. Bunce, John Thackray. *Fairy Tales: Their Origins and Meaning, With Some Account of Dwellers in Fairyland*. London, 1878.
12. Burenhalt, Goran (ed.). *The First Humans: Human Origins & History to 10,000 BC. (The Illustrated History of Humankind, Vol. 1).* New York: Harper Collins, 1993.
13. Bynre, Richard. "The Manual Skills and Cognition That Lie Behind Hominid Tool Use." *The Evolution of Thought: Evolutionary Origins of Great Ape Intelligence*. Ed. Anne E. Russon & David R. Begun. Cambridge: Cambridge University Press, 2004.
14. Call, Josep & Michael Tomasello. *Primate Cognition*. New York: Oxford University Press, 1997.
15. Childress, David Hatcher. *Technology of the Gods*. Kempton, IL: Adventures Unlimited Press, 2000.
16. Ciochon, Russell, John Olsen & Jamie James. *Other Origins: The Search for the Giant Ape in Human Prehistory*. New York: Bantam Books, 1990.
17. Ciochon, Russell. "The Ape That Was: Asian Fossils Reveal Humanity's Giant Cousin." *Natural History*, Nov. 1991. <http://www.uiowa.edu/~bioanth/giganto.html>
18. Cole, Sally J. *Legacy On Stone: Rock Art of the Colorado Plateau and Four Corners Region*. Boulder, CO: Johnson Books, 1990.
19. Cowen, Richard. *History of Life*. Malden, MA: Blackwell Science, 2000.
20. Cremo, Michael A. & Richard L. Thompson. *Forbidden Archaeology: The Hidden History of the Human Race*. Los Angeles: Bhaktivedanta, Book Publishing 1998.
21. Cunliffe, Barry. *The Ancient Celts*. New York: Penguin Putnam, 1997.
22. Dahinden, René, & Don Hunter. *Sasquatch/Bigfoot: The Search for North America's Incredible Creature*. 1973. New York: Firefly Books, 1993.
23. David, Rosalie. *Handbook to Life in Ancient Egypt*. New York: Facts on File, 1998.
24. De Beaune, Sophie A. "Prehistory and Cognition." *Current Anthropology*, Apr. 2004: 139-162.
25. "Fantastic Footprints." *Archaeology*, Sep./Oct. 2005: 12.
26. Farid, Shahina. "Excavating Catalhoyuk." *Archaeology Odyssey*, May/June 2005: 26-32.
27. Faulkner, Raymond O., trans. *The Ancient Egyptian Pyramid Texts*. New York: Oxford University Press, 1969.
28. Fischman, Josh. "Family Ties." *National Geographic*, Apr. 2005: 16-27.

29. "Geologic Time Online Edition." *General Interest Publications of the US Geological Survey*. Denver: USGS Information Services, 2000. 15 Aug. 2005. <http://pubs.usgs.gov/gip/geotime/>
30. Gilroy, Rex. "Aboriginal Traditions, Myths, and Legends." The Australian Yowie Research Center, 16 Mar. 2003. <http://www.theaustralianyowieresearchcenter.com/aboriginal_legends.html>
31. Gordon, Stan. "A Truly Mysterious UFO/Bigfoot Encounter." *The World's Best UFO Cases*. Ed. Dwight Connelly. Martinsville, IL: Bookseller, Inc., 2004.
32. Gore, Rick. "National Geographic Research and Exploration: New Find." *National Geographic*, Aug. 2002: n. pag.
33. Green, John. *Sasquatch: The Apes Among Us*. Seattle, WA: Hancock House, 1978.
34. Hancock, Graham. *Fingerprints of the Gods*. New York: Three Rivers Press, 1995.
35. Humphrey, Nicholas. "The Social Function of Intellect." *Growing Points in Ethology*. Ed. P.P.G. Bateson & R.A. Hinde. Cambridge: Cambridge University Press, 1976.
36. Hurley, Matthew. *The Alien Chronicles: Compelling Evidence for UFOs and Extraterrestrial Encounters in Art & Texts Since Ancient Times*. Chester, UK: Quester Publications, 2003.
37. "India Finds More 'Tsunami Gifts'." BBC News online, 27 Feb. 2005. <http://news.bbc.co.uk/2/hi/south_asia/4302115.stm>
38. Joose, Andy. "Tutankhamen's Perplexing Calcite Barque (& The Secret of the Sloppy Painter)." *Kmt: A Modern Journal of Ancient Egypt*, Spring 2005: 34-39.
39. Krantz, Grover. *Bigfoot Sasquatch Evidence*. Blaine, WA: Hancock House, 1999.
40. Lewin, Roger. "The Earliest 'Humans' Were More Like Apes." *Science*, 236 (1987): 1061-1063.
41. Marshack, Alexander. "An Ice Age Ancestor?" *National Geographic*, Oct. 1998: 478-481.
42. McKie, Robin. *Dawn of Man: The Story of Human Evolution*. New York: Dorling Kindersley, 2000.
43. Mitchell, Melanthia. "Scientists Detail Study of Kennewick Man." Yahoo! News, 11 Jul. 2005. <http://news.yahoo.com/s/ap/kennewick_man;_ylt=A86.I1xwrd9CbC4AwA1FeQoB;_ylu=X3oDMTBiMW04>

44. Moneymaker, Matt. "Deer Kills and Bigfoots." Bigfoot Field Researchers Organization, 2005. <http://www.bfro.net/AVEVID/MJM/deerkills.asp>
45. Mora, Rafael, et al. "The Archaeology of the Peninj 'ST Complex' (Lake Natron, Tanzania)." First Hominid Technology Workshop. Universitat Autonoma de Barcelona. Centre d'Estudis del Patrimoni Arqueològic de la Prehistòria, Bellaterra, Spain. Dec. 2001. Reprinted at http://www.bib.uab.es/pub/arqueologia/11349263n9p77.pdf.
46. Morwood, Mike, Thomas Sutikna & Richard Roberts. "The People Time Forgot." *Naitonal Geographic,* Apr. 2005: 2-15.
47. Noah, Sherna. "English Civil War Documents Unearthed." *The Scotsman* online edition (Dec. 13, 2004). <http://news.scotsman.com/latest.cfm?id=3882293>
48. Noble, David Grant. *Ancient Ruins of the Southwest: An Archaeological Guide.* Flagstaff, AZ: Northland Publishing, 2000.
49. O'Neil, Dennis. "The Primates: Apes." 2005. <http://anthro.palomar.edu/primate/prim_7.htm>
50. Pettifor, Eric. "From the Teeth of the Dragon: *Gigantopithecus blacki.*" *Selected Readings in Physical Anthropology,* Ed. Peggy Scully. Dubuque, IA: Kendall/Hunt Publishing, 2000. 143-149. <http://www.wynja.com/arch/gigantopithecus.html>
51. Pye, Lloyd. Starchild UK website, 28 Aug. 2005. <http://www.starchild-uk.com/>
52. Reeves, Nicholas. *Akhenaten: Egypt's False Prophet.* New York: Thames & Hudson, 2001.
53. Rennard, Andy. "Pre-Columbian and Early American Legends of Bigfoot-Like Beings." The Bigfoot Field Researchers Organization, 18 June 2000. <http://www.bfro.net/legends/>
54. Rice, John. "Bone Find in Mexico Baffles Archaeologists." *Fort Worth Star-Telegram,* 3 Oct. 2004, 20A.
55. Robertson, Dawna. "History on the Rocks: Hawaiian Petroglyphs." *TravelAge West,* Sep. 18, 2000: B1.
56. Schick, Kathy D. & Nicholas Toth. *Making Silent Stones Speak: Human Evolution and the Dawn of Technology.* New York: Simon & Schuster, 1993.
57. Shubinski, Raymond. "Lure of the Pleiades." *Astronomy,* Jan. 2005: 74-77.
58. Sighting Report. National UFO Reporting Center website, 3 Apr. 2005. <http://www.nuforc.org/webreports/015/S15938.html>

59. Skurdenis, Julie. "Focus on Archaeology: The Geoglyphs of Chile." *International Travel News*, Feb. 2001: 172.
60. Stanford, Craig B. "Chimpanzee Hunting Behavior and Human Evolution." *American Scientist*, 83 (1995): 256-261.
61. Stanford, Dennis. "Immigrants From the Other Side?" Center for the Study of the First Americans, n.d. <http://www.centerfirstamericans.com/cat.html?c=5>
62. Stringer, Chris. "Has Australia Backdated the Human Revolution?" *Antiquity*, 73 (1999): 876.
63. Stringer, Christopher & Robin McKie. *African Exodus: The Origins of Modern Humanity*. New York: Owl Books, 1996.
64. Tattersall, Ian. *The Fossil Trail: How We Know What We Think We Know About Human Evolution*. New York: Oxford University Press, 1995.
65. Vallee, Jacques. *Dimensions: A Casebook of Alien Contact*. New York: Ballantine Books, 1988.
66. Von Däniken, Erich. *Chariots of the Gods: Unsolved Mysteries of the Past*, 1970. Trans. Michael Heron. New York: Berkley Books, 1999.
67. Whoops and Knocks. Recorded 1974 by the Berry-Morehead expedition, Sierra Nevada Mountains, CA. <http://www.bfro.net/avevid/SierraSounds/Whoop.mp3 >
68. Wong, Kate. "The Littlest Human." *Scientific American*, Feb. 2005: 56-65.

LISA A. SHIEL founded the Michigan Upper Peninsula Bigfoot Organization (MUPBO) in 2005 to satisfy UP residents' intense interest in the subject. As a recognized expert on Bigfoot, Lisa has been interviewed by big-city newspapers, drive-time talk radio hosts, and TV reporters. She's a certified professional librarian with a master's degree in library science, as well as a certified field investigator for the Mutual UFO Network. As a fiction writer, Lisa developed the Human Origins Series—which includes the novels *The Hunt for Bigfoot* and *Lord of the Dead*, with a third book in the works. *Backyard Bigfoot*, her first foray into nonfiction, is the culmination of years of research into Bigfoot, human origins, UFOs, and ancient history.

If you found this book educational, entertaining, or intellectually stimulating and would like to acquire additional copies for youself or others, please contact your local bookstore, one of the online booksellers (such as Amazon.com), or visit the publisher's website at *www.SlipdownMountain.com.*

Please visit the Human Origins website at *Origins.SlipdownMountain.com* to read the Strange Origins Blog and check out the links to interesting news and information about hairy hominids, UFOs, archaeology, anthropology, and other related topics.

If you have comments or questions for Lisa concerning this book or any of the above topics, email her at *Lisa@SlipdownMountain.com.*

CPSIA information can be obtained at www.ICGtesting.com
Printed in the USA
BVOW05s0418040414

349578BV00003B/142/A

9 780974 655369